Evolution
third edition

Jay M. Savage
University of Southern California

Holt, Rinehart and Winston
New York Chicago San Francisco
Atlanta Dallas Montreal
Toronto London Sydney

Library of Congress Cataloging in Publication Data
Savage, Jay Mathers.
    Evolution.

    (Modern biology series)
    Includes bibliographies and index.
    1. Evolution.
QH366.2.S28   1977   575   76-26696
ISBN 0-03-089536-7

Printed in the United States of America
 8 9 0   090   9 8 7 6 5 4 3

Preface

The function of *Evolution,* Third Edition, is
to provide a basic understanding of the in-
conceivably complex but unifying theme of biological evolutionary change within the broad
framework of cosmic evolution. This book is
written in the strong conviction that the study
of evolution is fundamental to the understanding of any field of biology and forms a rich
area of investigation in its own right.

No serious biologist today doubts the
fact of evolution, the development of all living
organisms from previously existing types under the control of evolutionary processes. However, there have been and will continue to be
differences of opinion on *how* evolution takes
place, just as there are different ideas on the
exact processes in the formation of mountain
ranges, for example. Thus, although the fact

of evolution is amply clear, there are different theories regarding the significant processes that have brought about evolutionary change.

In this book we are not concerned with enumerating the so-called proofs of evolution. The fact of evolution is demonstrated on every side in all fields of biology and indeed forms the basic unifying principle in the study of living systems. We do not need to demonstrate the existence of mountain ranges. Rather, we will be concerned here with what is known about the process of evolution and with a survey of the several theories proposed to explain the process. In particular, I will develop the most generally accepted theory of evolution as supported by material evidences, observations, experiments, and theories of basic evolutionary forces. An understanding of the manner in which evolution develops should provide all the evidence necessary to convince men with open minds of the reality of this phenomenon. For those interested in greater detail and depth than can be provided by this necessarily succinct review, the Further Readings at the end of each chapter indicate sources for more in-depth study. In these lists I have emphasized books over short articles, simply because I myself often begin to read only a particular section of a book, but then I am led to explore other sections often with gratifying results.

While my own educational background and this book, generally, are out of the tradition of modern science, I have attempted to treat some aspects of evolutionary thought that as yet are not easily amenable to strictly scientific approaches or to the reconstructed logic they entail. In this era of excessive and destructive rationality and Western man's bias toward treating all creatures, including himself, as manipulable objects, human beings have tended to become insensitive to life, nature, and existence. This book attempts to restore some balance to the scene of evolution by countering the more extreme concepts of reductionism, mechanization, and rigid scientific materialism with alternate views. My own interest in science arose from the curiosity engendered by my first field experiences and the sense of awe, beauty, and joy produced within me by my encounters with the natural world. The task of each human individual today remains as it always has been: to integrate the two most important aspects (feelings and intellect) of each human personality into a functional and integrated whole, so that the beauties of both worlds of man—the external world of facts, ideas, and science and the inner world of feelings, imagination, and myth—may be savored by passing at will from one world to another, but knowing the difference.

The present book is unique among discussions of evolution at the college level in its emphasis on the three crucial unsolved problems in the understanding of evolutionary processes: (1) By what means do isolating mechanisms develop to prevent genetic exchange between related populations of organisms and lead to the origin of species? (2) What processes are responsible for the origin of major evolutionary changes above the species level? (3) What has been the pattern of the evolution of consciousness and what is its meaning to man? The final solutions to these problems may never be found, but I hope that some readers of this book will be challenged to undertake the study of them.

The preparation of the third edition involved substantial revision and inclusion of new materials. Because of its profound importance as a challenge to the understanding of evolution I have added a new chapter (11) tracing the origin and evolution of life. In addition the chapter (12) on the evolution of man has been extensively revised.

For those interested in further materials on the biological evolution of our species, I especially recommend the outstanding book *Human Evolution,* 2d ed., by Bernard Campbell (Chicago: Aldine Publishing Co., 1974). In cases of my own doubts in controversial matters I have followed Campbell's interpretations. I also especially recommend to all readers the introductory pages (pp. 1–97) of *The Myth of the Machine, Technics and the Human Condition* (New York: Harcourt Brace Jovanovich, 1966) for a stimulating and creative statement on the evolving of man's gifts of consciousness, written in an elegant and evocative style.

I wish to acknowledge my debt to my teachers at Stanford University, especially Rolf L. Bolin, Gordon F. Ferris, George S. Myers, and David G. Regnery, who shaped my early ideas on evolutionary theory; to my graduate students at the University of Southern California, who encourage me to constantly revise and reexamine my views; and to the undergraduates, who demand growth and development in themselves and their instructor. The latter have especially taught me to appreciate that we humans are in the grip of uncontrollable and inexplicable forces beyond our comprehension. It remains our choice how we deal with the forces of existence. Or stated another way:

As human beings we have the desire
To explain the unexplainable
To control the uncontrollable
To predict the unpredictable
To ignore the unignorable
And to force the unknowable future
     into concepts and models based on
     the selectively remembered past.
All to what end?
To avoid the pain and joy of existence.

I have faith that the readers of this book will not use their scientific knowledge to this latter end.

Los Angeles, California
November 1976   *J. M. S.*

Contents

Modern Biology Series **Consulting Editors**

James D. Ebert
Marine Biological Laboratory

Howard A. Schneiderman
University of California, Irvine

Berns	**Cells**
Delevoryas	**Plant Diversification,** *second edition*
Ebert/Sussex	**Interacting Systems in Development,** *second edition*
Ehrenfeld	**Biological Conservation**
Fingerman	**Animal Diversity,** *second edition*
Griffin/Novick	**Animal Structure and Function,** *second edition*
Levine	**Genetics,** *second edition*
Novikoff/Holtzman	**Cells and Organelles,** *second edition*
Odum	**Ecology,** *second edition*
Ray	**The Living Plant,** *second edition*
Savage	**Evolution,** *third edition*
Sistrom	**Microbial Life,** *second edition*
Van der Kloot	**Behavior**

part *I*

INTRODUCTION

chapter **1**

Evolution and Life

Diversity and unity are the two underlying themes that seem to characterize all life. There are approximately 5 million different species of organisms living today, and many millions of species that formerly existed on the earth are now extinct. Diversity in all phases of life activity is found in this vast array of life — ranging from simple viruses, through unicellular organisms, to such complex and diverse entities as whales and palm trees. Less apparent than diversity, but equally typical of living organisms, is unity in basic characteristics. This is recognized by most of us when we see, even dimly, a common similarity in all life. A full realization of this unity, however, in terms of the fundamental characteristics of reproduction and the transformation and utilization of energy, has been developed only recently and constitutes one of the major triumphs of science in the twentieth century.

Although these two themes of diversity and unity in life may seem to be antagonistic or mutually exclusive, they are reconciled through the concept of evolution. This principle occupies a central unifying position within the field of biology because it explains how diversity and unity may be characteristic of living systems at the same time. It is the most significant concept developed in the study of living organisms, provides explanations for myriad biological processes, and pervades every branch of biology — from biochemistry and physiology to ecology and morphology. Indeed, evolutionary thought influences every field of knowledge.

Many astronomers and astrophysicists assure us that our universe was born about 15 billion years ago when a thermonuclear blast vaporized the original ball of matter. Built into the universe from that moment, when time began, was the cosmic imperative of evolutionary change. Within minutes after the beginning, the atomic elements were formed as countless ultracosmos of matter and energy. Some 3 billion years later, about 12 billion years ago, the gradual cooling processes led to condensations of the basic particles into swirling clouds of gases. About 6 billion year ago one of these clouds had become sufficiently cooled and condensed into discrete units to form our solar system of nine planets circling our sun. The evolution of the earth as a home for life was thus begun. By about 4.5 billion years ago conditions on the earth's surface permitted the beginning of chemical evolution of atomic elements in new and more complex combinations to create the opportunities for the origin of life from nonliving chemical systems about 3.5 billion years ago. Organic evolution, the diversification of living systems, began at that time and continued through myriad changes to new evolutionary levels associated finally in man with cultural and psychic (consciousness and motivation) components, emergents from the base of our biological background.

As may be seen from the accompanying diagram (Figure 1-1), each stage in cosmic evolution overlaps the next proceeding stage. Gradually through time the earlier stages lose importance.

On the other hand, other influencial astronomers and astrophysicists, who agree with the first group regarding the age of our solar system and the earth, tell us that while the universe is expanding, the expansion is not due to an explosion. The galaxies rush apart because new matter is invading unoccupied space. This matter coalesces into new galaxies; so there was no beginning of our universe nor will there be an end—it will, like life itself, go on creating itself eternally.

The term "evolution" can thus be used in a variety of ways. The planet earth as we know it today is the result of historical changes or *geologic* evolution. The historical stages leading to the rise of human civ-

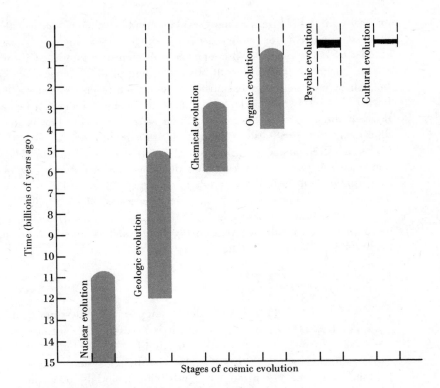

Figure 1-1 *The evolutionary imperative.*

ilizations may be spoken of as *cultural* evolution. And we may even refer to the evolution of airplanes, automobiles, radios, telephones, and ideas, although each of these forms a special case of cultural evolution. Even though these various kinds of evolution resemble one another in having the characteristic of a sequence of developmental stages, none, because of the unique qualities of living organisms, is precisely the same as organic evolution.

The present book is restricted to the analysis of that portion of cosmic evolution responsible for the unity and diversity of life. Hereafter, "evolution" will refer only to organic evolution, except when a phase of nonorganic evolution is clearly denoted by an appropriate modifier.

In a broad sense, biological evolution involves two different kinds of development. Each individual organism undergoes a developmental process from its time of origin until death. Every human being, for example, begins life as a single, small cell, and then passes through a series of increasingly complex changes, culminating in a fully developed, coordinated, multicellular adult. The evolution of the individual is called *ontogeny,* and because the most rapid period of change occurs during early development, or embryogenesis, ontogenetic changes are usually studied by embryologists. Although the evolutionist is interested in ontogeny, his primary concern is with a second kind of development. We know that all groups of living organisms — populations, species, genera, families, orders, classes, phyla, and kingdoms — have undergone changes through the course of time and that, as a result, living forms are the descendants of previously existing and rather different ancestors. The historical development of groups of organisms is referred to as *phylogeny,* and it is with phylogenetic evolution that most modern students of evolution are concerned. It is the result of phylogenetic evolution that all organisms exhibit unity in basic biological processes, since all the diverse lines of evolution are descendants of common ancestors. Diversity originates, then, as a result of phylogenetic change and is superimposed upon the unity provided by common ancestry.

THE NATURE OF LIFE

In order to appreciate fully the nature of evolution, it is necessary to consider the material basis of life. No one, generally applicable definition of life is possible, because living systems appear to be extremely complex organizations of nonliving materials, operating in accordance with the same physicochemical principles evident in the functioning of the inorganic world. Life is particularly difficult to define because it is a dynamic, continuously changing process of unimaginable complexity, harnessing the principles of physics and chemistry but by its nature transcending them. This complexity is reflected in the organization of living systems, which is much greater than that of nonliving ones. We know that all substances, living and nonliving, are formed of certain basic units called *atoms.* Atoms are combined in specific ways to form various kinds of *molecules.* All nonliving systems are composed of groups of atoms forming one kind of molecule (substances) or of several kinds of molecules (mixtures). Living systems are mixtures of very large and complex molecules functioning together in a coordinated manner. The very large and most complex nonliving molecules (*proteins*) approach in size and chemical activity the very smallest of living systems (*viruses*). Above the virus level, living material is organized into units that are called *cells.* These units are made up of a great many different kinds of complex molecules. Individual organisms are composed of one to a great many cells. Above the individual level are additional organizational units of increasing complexity: the *population,* the *species,* the *community,* and finally, the *ecosystem.* The spectrum of organization is illustrated in Table 1-1. It is important to recognize that just as every living thing is

Table 1-1   *Spectrum of Organizational Complexity*
(Each unit is made up of the units immediately below.)

	Unit	Composition
Most complex	Ecosystem	Community and nonliving environment
	Community	Several to many interacting species
	Species	One to several genetically similar populations
	Population	Several to many genetically similar individuals
	Individual	One to many cells
	Cell	Organized living material under regulatory control of genetic materials; complex colloidal mixture of carbohydrate, lipid, and protein molecules, in addition to other substances in water (Figure 1-2)
		Organized subcellular living material ranging in composition from a single large protein molecule to several protein molecules, containing DNA or RNA: the viruses
Simplest	Molecule	Two to many atoms
	Atom	Two to many fundamental particles (protons, neutrons, electrons) bound together by energy

Increasing complexity — LIVING SYSTEMS — NONLIVING SYSTEMS

composed of the basic physical units — atoms and molecules — each individual organism exists as part of a particular population, species, community, and ecosystem.

All living creatures are composed of a peculiar combination of substances, all of which occur in the nonliving world. The chemistry of the principal constituents of protoplasm is in large part responsible for the attributes of life. The most important components of all living systems are water (H_2O) and large molecules containing carbon atoms: the carbohydrates, lipids (fats), proteins, and nucleic acids. Proteins also share characters in common because of the presence of nitrogen and often sulfur in their structure. Complex proteinaceous catalysts, called *enzymes*, initiate and regulate most of the chemical activities of living systems. Finally, all the many different kinds of carbohydrates, lipids, proteins, nucleic acids, and enzymes, and small amounts of a great many inorganic materials, are organized into a colloidal mixture in the water. This physical arrangement facilitates complex chemical activity by increasing the

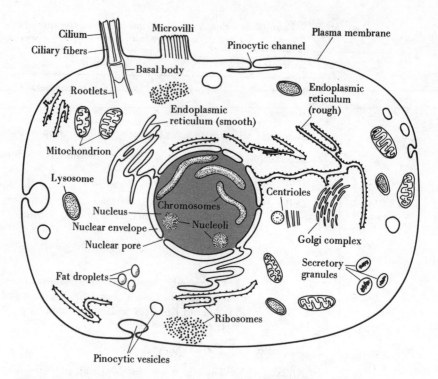

Figure 1-2 *Simplified diagrammatic representation of a typical cell. The various structures and organelles shown are not found in every cell but indicate the spectrum of differentiation in structure. The illustrated features function in the exceedingly complex cytoplasmic matrix, a rich collodial mixture of macromolecules, smaller organic molecules, and ions. DNA replication takes place in the nucleus; RNA transcription from the DNA occurs in the nucleus; protein synthesis, the translation of the RNA encoded messages into polypeptide chains, occurs in the ribosomes.*

number of interactions between the many kinds of molecules. That life manifests much greater complexity than any nonliving system is underscored when we realize that even the simplest cell is composed of thousands of different kinds of molecules operating in a coordinated fashion.

Special comment must be made concerning the extremely small organisms called *viruses*. These forms are similar to large protein molecules. All are made up of one to several protein molecules containing deoxyribonucleic acid (DNA) or ribonucleic acid (RNA). Their structure is without evidence of cellular organization, and although most biologists generally regard them as microorganisms, for a long time they were often regarded as peculiar nonliving systems. Most scientists today agree that viruses are bits of genetic material, roughly equal to fragments of the chromosomes of cellular forms. Significantly, viruses resemble very closely any proposed intermediate between nonliving molecules and living material. Whatever their status, the viruses show that there is no clear-cut break in the continuum from nonliving to living systems.

The complexity of the physicochemical organization of life makes possible a number of fundamental processes that occur in a regulated, organized fashion in all living systems. The first of these features is the product of the chemical organization and enzymatic activities of living material. All life has the ability to take nonliving materials and to convert them into part of the living system. Each one of us converts the nonliving foods from his daily meals into a living human being, and all other living organisms display this same capacity. Associated with this characteristic is the further ability to break down certain portions of the organism's own living substance in order to release energy for life's activities. *Conversion* and *energy production* are spoken of collectively as *autosynthesis* and are typical of all living systems from viruses to men. Viruses, however, are unique in that they display these features only when they are within the cells of some other living organism. No nonliving system exhibits the attributes of autosynthesis.

A second peculiar feature of life is its capacity for *reproduction*. Each type of living organism is able to produce new individuals that are essentially duplicates of the parent or parents. All human beings reproduce human beings, not snakes, sunflowers, or viruses. The processes of biological reproduction are ultimately under the control of special enzymatic regulators produced by the activity of fundamental hereditary units called *genes*. The genes are self-duplicating units and are basically responsible for the self-duplicating feature of biological reproduction. The genes are currently thought to be specialized chemical units of the DNA molecules in the chromosomes. Differences in the chemical structure of the DNA molecules are translated into the production of different enzyme systems. The differences in the sum total of all enzymatic activities, especially because these regulate ontogenetic development, are responsible for the differences between various kinds of living organisms. A lion is not an oak tree, for example, principally because a lion has an enzyme control system greatly different from that of an oak. The enzyme systems differ because the hereditary materials (genes) are different in the two kinds of organisms. The special kind of self-duplicating reproduction made possible by the genes is called *autocatalysis*, and it is typical of all living organisms. Viruses alone among living creatures are autocatalytic only when they are within the cell of other organisms. The viruses are primarily composed of DNA or RNA surrounded by a proteinaceous coat, and an individual virus consists of from one to several genes. No nonliving system is capable of the precise autocatalytic reproduction found in living organisms.

The third major characteristic of all living systems is the capacity for *adaptation,* or the regulated adjustment to a changing world. Every organism has some capacity for maintaining life's activities through modifications that adjust to sudden changes in the surrounding environment. For example, if the air temperature outside a human being's body goes up, the person will make a series of adjustments, including perspiration, to keep the body temperature at a constant level. Short-term internally regulated responses to external changes or stimuli are expressions of *adaptability*. Such stimulus-response adaptations adjust the individual organism to the flux of external circumstance. All organisms are, in addition, products of long-term adaptation to a gradually changing world. Each living organism is descended from ancestors who were not adapted in exactly the same way to the general features of the environment. Long-term adaptation is called *evolution* and is as typical of living systems as autosynthesis, autocatalysis, and adaptability. In contrast to these characteristics, evolution is a response by a population rather than by an individual. The initial change is internal in origin and is regulated by the external environment in the form of natural selection.

THE MOLECULAR BASIS OF LIFE

Within all living systems many basic structures of the organism are formed by proteins, and their basic activities are regulated with precision by specific catalytic proteins—the enzymes. Any understanding of the nature of life at the molecular level must obviously center on the nature of proteins and protein production.

Proteins are composed of long chains of amino acid molecules connected by a special kind of chemical bond (the peptide bond) and are consequently called polypeptide chains. Twenty different amino acids are found in proteins, and their number and arrangement determine the structure and activity of the protein. A given protein usually is composed of 100–300 different amino acid molecules, ordered in precise fashion into a polypeptide chain. For example I—IV—IV—XX—XV— . . . might represent the beginning of a polypeptide chain with each different amino acid indicated by a roman numeral. A protein with an initial sequence of I—IV—X—XX—XV— . . . will be different from the first one. Because the amino acids may be combined in an almost limitless number of possible sequences, it is not surprising that thousands of different proteins are known and that a single cell may contain many different specific enzymes. The discovery of the mechanism by which the sequences of amino acids are ordered in a living system during protein synthesis is among the great advances in twentieth-century biology. The process is complicated, and only the essential features as they relate to evolutionary thought are outlined here.

It is now well established that the production of proteins in a cell

is ultimately, if indirectly, controlled by deoxyribonucleic acid (DNA).
In some primitive organisms (bacteria), only a single chain of DNA is
present; in more highly evolved groups (mammals, higher plants) extremely long DNA chains form the axis of large nuclear structures—the
chromosomes. A single DNA molecule (Figure 1-3) consists of a very
long, paired series of repeating groups of phosphate and sugar, connected
at regular intervals by nitrogenous bases. The sugar-phosphate units form
a double helix, whereas the nitrogenous bases—one attached to each helix
strand and connected to the complementary nitrogenous base by a weak
chemical bond—form a series of "rungs" between helices. The nitrogenous
bases are the key elements in this discussion. Four kinds occur in DNA
molecules: adenine (A), cytosine (C), guanine (G), and thymine (T).
A single DNA chain may have as many as 200,000 bases along its axis,

P — phosphate
S — sugar
A — adenine
T — thymine
G — guanine
C — cytosine

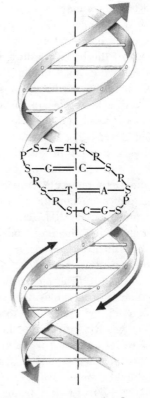

*Figure 1-3 Diagrammatic representation of a portion of a deoxyribonucleic
acid (DNA) molecule to indicate the double helix strands and the connecting
nitrogenous base pairs. The helix strands are formed by alternating sugar and
phosphate group; the connection by complementary bases—adenine, cytosine,
guanine, and thymine—as shown for a section of the molecule.*

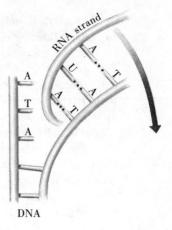

Figure 1-4 Diagrammatic illustration of production of single-strand ribonucleic acid (RNA) through action of DNA: transcription. Nitrogenous bases are adenine (A), cytosine (C), guanine (G), and uracil (U).

and the bases may be in any order. One other feature of the bases is also critical. Chemical bonds form between A and T or C and G, but not in any other combination. Thus, the pairs of nitrogenous bases running the length of the helix are always complementary to those on the other helix.

DNA is not capable of manufacturing proteins directly, but orders their production through a series of ribonucleic acid (RNA) molecules. RNA is a molecule similar to DNA with a series of nitrogenous bases, three of which (A, C, and G) are the same as in DNA, and a fourth, uracil (U), is complementary to A. RNA occurs in three functional forms in living cells, all produced by DNA. Protein synthesis is initiated (Figure 1-4) by the breaking of the weak chemical bond between the base pairs of the DNA molecule. Apparently the bases on the DNA helix strand can then pull complementary bases out of the surrounding material and form a complementary RNA chain. This *messenger RNA* becomes a long chain that embodies the basic plan and order of amino acid molecules for a protein molecule. *Transfer RNA* passes out into the cell and, depending upon its nitrogenous base series, will pick up an amino acid molecule (Figure 1-5). *Ribosomal RNA* combines with proteins to form ribosomes, the site of protein synthesis. The transfer RNA loaded with a specific amino acid will match up to a complementary series of three bases on the messenger RNA strand as the ribosome moves along the latter. Each three base series or triplet of the messenger RNA attracts a loaded transfer RNA that complements it. Each transfer RNA loses its amino acid molecule in turn, as the amino acid becomes attached to the growing polypeptide chain (Figure 1-6). The key feature of this process is the determination of protein specificity by the DNA molecule. The special characteristics of the DNA molecules within each living organism provide a specific blueprint for protein and enzyme synthesis (the process is summarized in Figure 1-7). Evolution ultimately results from structural changes in DNA.

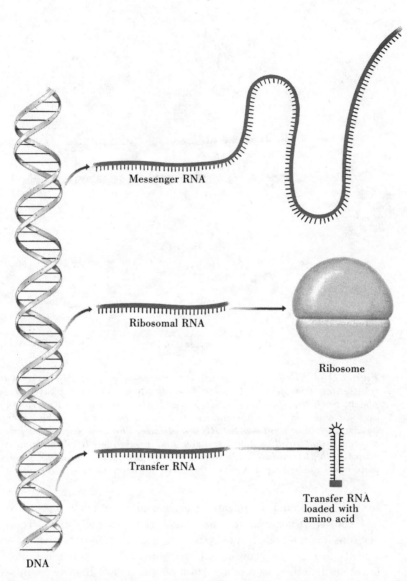

Figure 1-5 Diagrammatic representation of the production of the three kinds of RNA that cooperate in protein synthesis.

In addition to the unifying processes and characteristics common to all living systems, each form of life exhibits a series of nonreducible features that appear during their life history. These features may be generalized as follows: *organizationals, developmentals, regulationals, functionals,* and *emergents.* For example, each kind of organism has a basic and rec-

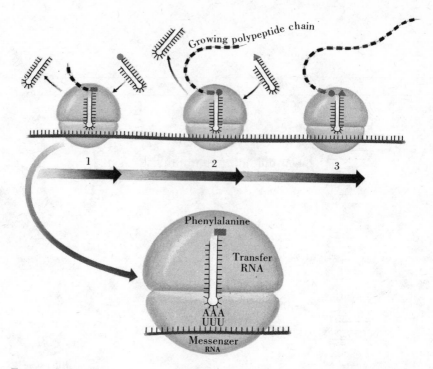

Figure 1-6 *Diagram representing the process of protein synthesis in cell: translation. Transfer RNA molecules pick up specific amino acids from cytoplasm. Each transfer RNA molecule adds its amino acid to a growing polypeptide chain in the ribosome in an order determined by the order of the base sequence of a messenger RNA molecule. The base sequence of the DNA molecule that produced the transfer and messenger RNA molecules determines both the amino acids picked up by the former and the sequence in which the messenger RNA orders their addition to the polypeptide chain.*

ognizable form and organization; each kind of organism has a distinctive and consistent pattern of growth and change; each kind of organism develops an integrated system for regulating its own activities and for contributing to the growth and function of its populational unit; each organism functions as a totality in a unique and distinctive fashion within its biotic community and ecosystem; and each kind of organism lies on a spectrum of emergent properties or qualities that harness those features which appeared earlier in evolution but transcend them. The emergents are associated with historically new life-styles and are the result of evolutionary change leading to a qualitatively new relation with the world. For example, the shift from unicellular organisms to the cooperatively integrated and functionally operational multicellular system opened up a whole series of new possibilities for life that cannot be explained on the basis of our knowledge of unicellular forms.

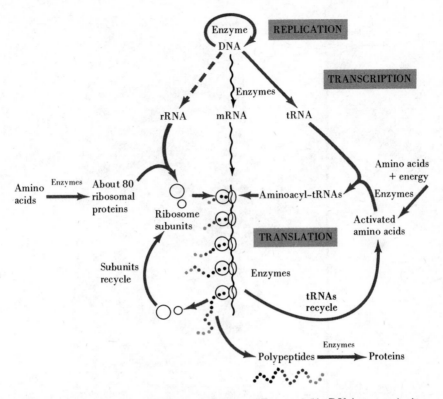

Figure 1-7   Summary of DNA replication (Figure 2-5), DNA transcription to RNA (Figures 1-3, 1-4), and translation of the encoded message on the RNA into polypeptide chains at the ribosomes (Figure 1-6). rRNA = ribosomal RNA, mRNA = messenger RNA, and tRNA = transfer RNA. The synthesis of a specific polypeptide or protein molecule requires the intervention of 100 or more different enzymes, each of which is also the product of the transcription-translation process. The enzymes involved are each specific to a single step in the process.

Life harnesses the principles of physics and chemistry but transcends them to produce irreducible higher biological principles. Contrary to the hopes of the most confirmed reductionists and mechanists, life cannot be reduced to, or explained simply in terms of, energetic, atomic, molecular, or mechanistic models. The reductionist-mechanistic approach provides enormous understanding of individual physicochemical units and mechanisms within living systems, but when extrapolated upward through the hierarchy of biological organization (Figure 1-8), it lacks explanatory power and ignores the most significant differentia of existence, the autonomous, goal-seeking, self-organizing, self-regulating, and time-patterned features of life.

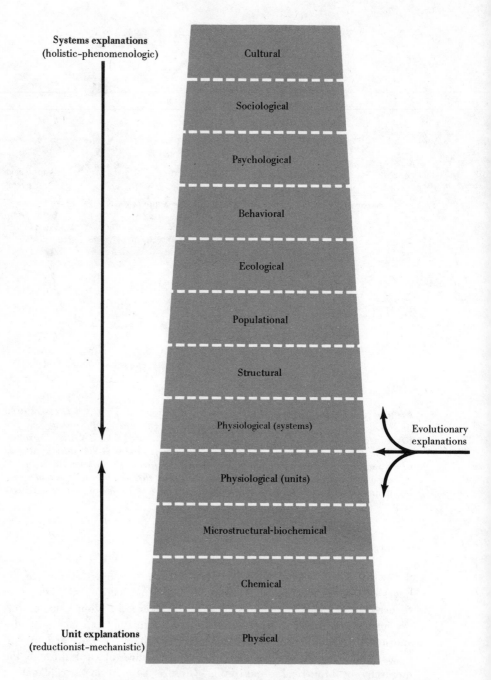

Systems explanations
(holistic–phenomenologic)

Cultural

Sociological

Psychological

Behavioral

Ecological

Populational

Structural

Physiological (systems)

Evolutionary
explanations

Physiological (units)

Microstructural-biochemical

Chemical

Unit explanations
(reductionist-mechanistic)

Physical

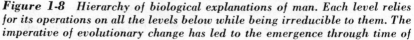

Figure 1-8   Hierarchy of biological explanations of man. Each level relies for its operations on all the levels below while being irreducible to them. The imperative of evolutionary change has led to the emergence through time of

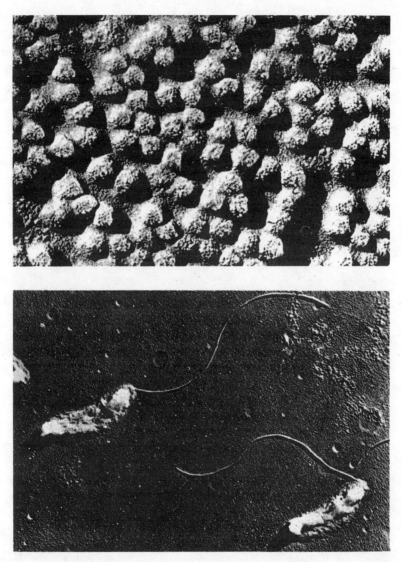

Figure 1-9 *Electronmicrographs of* (upper) *wound-tumor virus,* ×100,000, *and* (lower) *a bacterium,* Vibrio, ×5000. *(Courtesy of R. F. Bils)*

each higher level with its principles transcending principles operating at the level just below. Any level forms an interesting and valid sphere for biological study and explanation, since the levels are defined and the answers limited operationally by how questions and problems are approached. Attempts to extrapolate causal explanations from a lower to a higher level in the hierarchy tend to be simplistic and often false and destructive. Analytic descent from higher levels to lower levels often produces powerful general explanatory statements, but these ignore the most significant aspects of the higher level phenomenon and have less depth or compactness.

DIVERSITY AND HISTORY OF LIFE The complexity of living material and the significance of evolutionary adaptation as a fundamental feature of life are demonstrated in the vast differences among the principal groups of living organisms. Evolutionary and biochemical studies indicate that all kinds of organisms—about 2 million different living species, and the millions of extinct types—are descendants of a common early form of life (Figures 1-9, 1-10, and 1-11). The diversity of living organisms has been produced through adaptation to the world's many environments. Fundamentally, five major groups (kingdoms) of living creatures may be recognized, as enumerated below.

PRIMITIVE FORMS
(Each subgroup within the major divisions
equals a phylum in formal classification)

A. PROKARYOTES
No organized nucleus; no nuclear membrane; one linkage group
1. Monerans
   Viruses
   Bacteria
   Blue-green algae

B. PRIMITIVE EUKARYOTES
An organized nucleus; a nuclear membrane;
two or more linkage groups (chromosomes)
1. Protistans
   Red algae
   Cryptomonads and dinoflagellates
   Yellow-green algae, green-brown algae, and diatoms
   Brown algae
   Zooflagellates
   Rhizopods
   Sponges
   Ciliates
   Sporozoans
   Euglenoids
   Green algae
2. Fungi
   Slime molds
   Fungi

C. ADVANCED EUKARYOTES
3. Animals (well-organized multicellular types grouped into 20–25 phyla)
4. Plants (well-organized multicellular types, usually capable of photosynthesis, placed in two phyla)

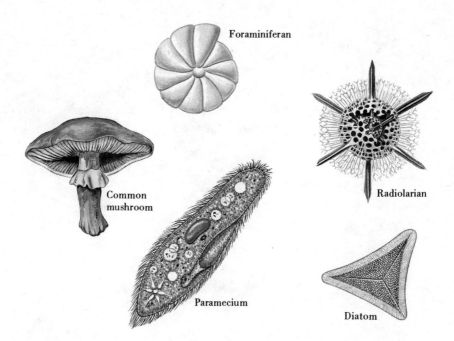

Foraminiferan

Radiolarian

Common
mushroom

Paramecium

Diatom

Figure 1-10 *Typical examples of the Kingdom Protista, all approximately ×75, and Fungi.*

The details of the relations of the monerans and protistans are not clearly understood, but it is evident that animals are derived from some kind of zooflagellate ancestor and that plants are closely related to the more primitive green algae.

Although the variety of living organisms known to exist on earth at the present time is impressive, we must remember that a study of the earth's history reveals thousands of kinds of extinct organisms. Fossils are the remains of dead plants and animals preserved in the record of the rocks, and from these remnants we are able to reconstruct in broad outline the history of living organisms. Since fossils from the oldest rocks are rather rare, the further back in time we go the less evidence we have from fossilized remains. On the basis of specialized dating techniques developed by astronomers, physicists, and geologists, we believe that the earth was formed about 6 billion years ago. Biochemical theory indicates that life could have first evolved from nonliving materials under the conditions probably present on the earth about 3.5 billion years ago. The first macrofossils are known from rocks formed about 1 billion years in the past; and some microfossils are known from 3.1 billions of years ago, but we have very little in the record of two-thirds of the history of life. Fossils

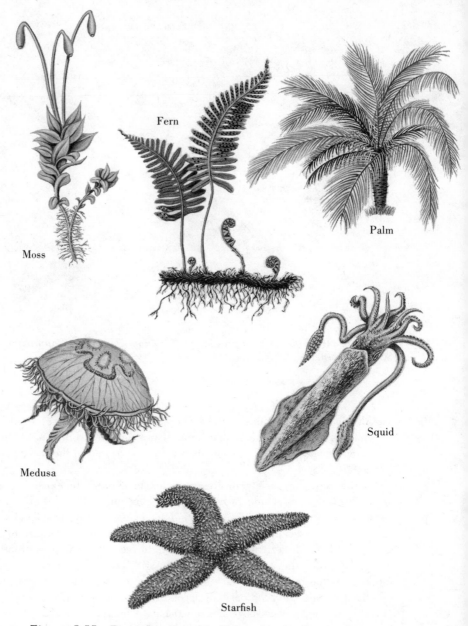

Figure 1-11   Examples of multicellular plants and animals.

and other remains of multicellular organisms become abundant in rocks of 600 million years ago. It is obvious from these approximate dates and from Table 1-2, which shows the main course of the earth's history, that ample time for evolutionary diversification has passed since life first appeared on the planet earth. Perhaps equally impressive is the maintenance of the basic unity of living systems through the vicissitudes of 3 billion years of evolutionary change.

In summary, life is a unique complex combination of nonliving materials—autonomous, self-organizing, self-regulated, and time-patterned —that expresses itself in a recognizable pattern of chemical reactivity (autosynthesis), reproduction (autocatalysis), and adaptation. In addition, each form of life exhibits a distinctive set of organizational, developmental, regulational, functional, and emergent qualities. All these properties are evoked by the structure of DNA, although not determined by it. The process of long-term adaptation has produced, over the 3.5-billion-year period since living material originated, a diversity of living creatures, many now extinct, and a spectrum of newly emergent life-world relationships. This diversity and its emergence confirms and exemplifies evolutionary change as a fundamental characteristic of life, but at the same time they have not eliminated the essential unity of all living systems.

HISTORICAL PERSPECTIVES   The idea of evolution, in common with most great human concepts, is not entirely of recent origin. The essence of the idea appears in Greek writings (600 B.C.) and probably occurred in many others throughout human history, although it was never generally accepted. Insofar as modern biology is concerned, the first clear recognition and demonstration of the fact of evolution was made by the French naturalist Jean Baptiste Lamarck (1744–1829). Lamarck's earliest paper on evolution appeared in 1801, but his principal theory was elucidated from 1815 to 1822. Lamarck brilliantly discerned that all life is the product of evolutionary change, that evolution resulted in the taking on of new adaptations to the environment, and that the diversity of life was the result of adaptation. Unfortunately, Lamarck developed a theory of evolution that does not stand up under investigation. Moreover, his ideas were generally rejected by other biologists, and evolution was ignored for nearly 45 years.

Lamarck's recognition of evolution as a fact and of progressive adaptation as a general theme in the history of life is usually overlooked by most biologists because his theory of the causes of evolution was faulty. Basically the theory consisted of three points: (1) Every considerable or permanent change in the environment of any organism produces a change in the organism's needs. (2) New or enlarged structures appear because of the "inner want" of the organism to meet these needs. (3) Structures are acquired, enlarged, reduced, or lost through use and disuse, and

Table 1-2   *The History of Life*

Time (in millions of years since beginning of epoch)	Eras	Epochs	Time (as percent since origin of life)	MAJOR EVOLUTIONARY EVENTS — Type of Evolution	First Appearance	Peak of Radiation	Time (on a 24-hour scale since origin of life)
3	Cenozoic	Pleistocene	99.9	Cultural / Psychic	Man / Human precursors	Flowering plants / Insects / Bony fishes	11:59 P.M.
12		Pliocene					
25		Miocene					
36		Oligocene				Mammals	
58		Eocene					
65		Paleocene	98				11:31 P.M.
130	Mesozoic	Cretaceous	96		Mammals		
180		Jurassic	95		Flowering plants	Reptiles	10:19 P.M.
230		Triassic	93				

Million years	Era	Period	No.	Mode	Events	Dominant forms	Clock time
280	Paleozoic	Permian	92		Gymnosperms		
310		Pennsylvanian	91		Reptiles	Amphibians	
340		Mississippian	90	Organic			
400		Devonian	89		Amphibians	Tree ferns	9:22 P.M.
450		Silurian	87		Insects / Land plants		8:53 P.M.
500		Ordovician	86		Fishes / Algae		
600		Cambrian	83		All major invertebrate phyla		7:55 P.M.
3500	Precambrian	Proterozoic	71	Chemical	First macrofossils / First life (3500)		5:25 P.M.
		Archeozoic	0	Geologic / Nuclear			12:00 MIDNIGHT
					Origin of earth (6000)		
12,000					Origin of universe (12,000)		

these changes are inherited by subsequent generations. The first statement is a fact. The second is not testable by scientific methods; the final conclusion is. In theory at least, the concept of the *inheritance of acquired characteristics* seems logical, particularly when we consider our legal and social systems, in which things (particularly money, objects, and land) acquired by one generation are passed on to the next. In addition, to a certain extent all individuals do adapt their bodies to the environment, as is seen by the degree of muscular development in the arms and shoulders of a man whose life is spent lifting heavy objects.

The essence of the Lamarckian theory is summarized in Figure 1-12, showing Lamarck's giraffe. The figure also shows a giraffe evolving according to Darwin's theory of evolution, which is discussed below. While the hypothesis is a reasonable speculation, it fails when subjected to scientific test. The objections to it are several: there is no known way by which somatic cells may pass characteristics to reproductive cells; experiments show no inheritance of acquired characteristics (the beach enthusiast's children are not born with a suntan); many adaptations found in organisms are not of a type that could be acquired (can an animal practice being purple?); and some acquired characteristics cannot possibly be passed on (neuter sex in worker bees). Lamarck was correct in his insight as to the significance of evolution, but he failed to muster a satisfactory theory to explain how it occurred.

Although the idea that direct action of the environment induced inherited changes is often credited to Lamarck as a major aspect of his theory, he specifically rejected the possibility of such environmental effects. Later, after the appearance of Darwin's work, a group of scientists, who came to be called neo-Lamarckians, championed this concept, on the basis of indirect evidence, as the principal mechanism of evolution, but the absence of experimental tests in support of this hypothesis caused it also to be abandoned as an explanation.

The world was to wait almost half a century before the genius of Charles R. Darwin (1809–1882) provided the key theory of evolution and at the same time converted the scientific and intellectual worlds to acceptance of the fact of evolution. Darwin's masterpiece entitled *On the Origin of Species by Means of Natural Selection* appeared on November 24, 1859. It became a best seller overnight and changed man's thinking forever. The time was ripe, and the impact both on scientists and on others—in fact everyone—was profound. We cannot consider here all the ramifications of Darwinism, but interested readers are directed to the list of references at the end of the chapter. The biologically important aspects of *On the Origin of Species* include three points: the recognition of evolution as a fact; the presentation of data demonstrating the fact; and the development of a theory of how evolution occurred. Darwin attributed evolutionary

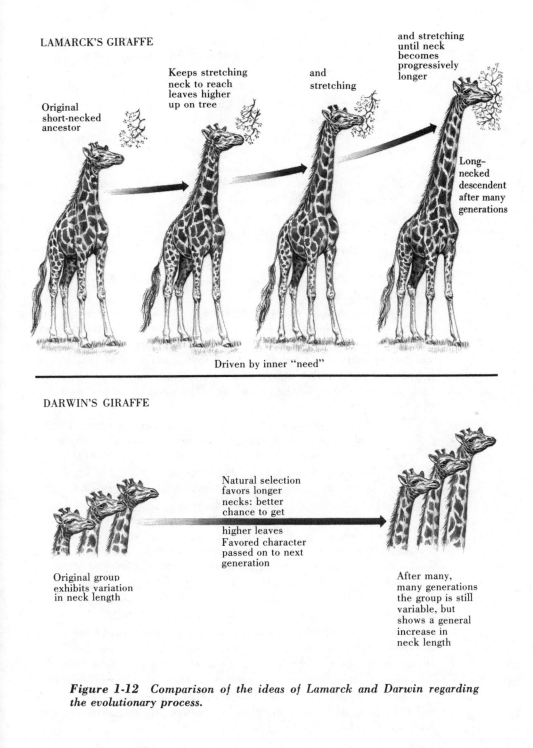

LAMARCK'S GIRAFFE

Original short-necked ancestor

Keeps stretching neck to reach leaves higher up on tree

and stretching

and stretching until neck becomes progressively longer

Long-necked descendent after many generations

Driven by inner "need"

DARWIN'S GIRAFFE

Natural selection favors longer necks: better chance to get

higher leaves Favored character passed on to next generation

Original group exhibits variation in neck length

After many, many generations the group is still variable, but shows a general increase in neck length

Figure 1-12   Comparison of the ideas of Lamarck and Darwin regarding the evolutionary process.

change to several forces, but the prime force was natural selection. His theory is based on two observations and two basic conclusions.

Facts
1. All organisms exhibit variability (look around any classroom).
2. All organisms reproduce many more offspring than survive (the North Atlantic female cod lays 85 million eggs at once).

Conclusions
1. The environment selects those individuals best fitted to survive, while individual variants less well fitted fail to reproduce (natural selection).
2. The characteristics thus favored by selection are passed on to the next generation.

Figure 1-12 also shows Darwin's giraffes, which evolved in a different manner from Lamarck's.

The most serious weakness of Darwin's explanation stemmed from his lack of knowledge about heredity. He was sure that differences were inherited, but *how* this occurred was unknown. Darwin died before the secret of heredity was penetrated. It was uncovered by a Moravian monk, J. Gregor Mendel (1822–1884), but his work received no attention until 1900, so the fruits of his labor had no effect on evolutionary theory for 35 years.

Following the rediscovery of Mendel's principles, there began a period of rapid acceleration in knowledge of heredity leading to the science of genetics. Unfortunately, some early geneticists discarded almost all Darwin's ideas because, in that egoism so typical of man, they knew something he did not know. It even became fashionable to speak of the death of Darwinism.

During the years since 1920, however, it has gradually become clear that modern genetics provides the final large piece in the jigsaw puzzle of evolutionary theory. Instead of disproval of Darwin's ideas, a cross-fertilization has occurred to produce a generally satisfactory theory of evolution based on the interaction of heredity and natural selection. It is this "synthetic theory" that forms the basis for the following discussion of evolutionary processes in succeeding chapters of this book.

Equally important to the future of evolutionary thought are the discoveries of the preceding 15 years that have given biologists insight into the chemical nature of the hereditary materials and into the relationships between the latter and protein synthesis. These discoveries and newly developed concepts offer unexplored opportunities for studies on the molecular basis of evolutionary change and may provide unexpected ramifications of evolutionary thought.

FURTHER READING

Calvin, M., "Chemical evolution," *American Scientist*, vol. 63 (1975), pp. 169–177.

Eiseley, L., *Darwin's Century*. New York: Doubleday, 1958.

Gorney, R., *The Human Agenda*. New York: Simon and Schuster, 1968.

Irvine, W., *Apes, Angels and Victorians*. New York: McGraw-Hill, 1955.

Loewy, A., and P. Siekevitz, *Cell Structure and Function*, 2d ed. New York: Holt, Rinehart and Winston, 1969.

Oparin, A. I., *Genesis and Evolutionary Development of Life*. New York: Academic Press, 1968.

Polanyi, M., "Life's irreducible structure," *Science*, vol. 160 (1968), pp. 1308–1312.

part **II**

THE FUNDAMENTAL EVOLUTIONARY PROCESS

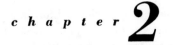

chapter 2

The Genetic Basis of Evolution

An essential key to the development of the modern theory of evolution lies in the discoveries emanating from the experiments carried out by Mendel in the 1860s. When Mendel's basic ideas were rediscovered in 1900, they produced a revolution in biological thought comparable to the impact of Darwin's theory of natural selection or, in our own time, to the effect of the concepts of molecular biology. Rediscovery of Mendel's principles led to the rapid, explosive growth of the field of genetics and established the basis for unraveling the secret of biological reproduction and heredity. A leading figure in developing our current knowledge of the patterns and mechanisms of biological inheritance was Thomas Hunt Morgan (1866–1945), who, with a great many associates, worked first at Columbia University and then at the California Institute of Technology. By the early 1940s, these researchers and a host of other geneticists had established a sound basis for explaining the fundamental patterns of biological inheritance. Included were the concepts of independent segregation, random assortment, the genetic factor (gene), the chromosomal theory of heredity, chromosomal regulation of sex determination, linkage, crossing-over, multiple gene control of development, and mutation.

Almost from the outset of genetic study, scientists began to speculate about the chemical nature and manner of action of genetic material. No real progress was made in this search until 1944, when O. T. Avery, C. M. MacLeod, and M. McCarty demonstrated that deoxyribonucleic acid (DNA), long known to be localized in the chromosomes, is the genetic material. Subsequently, in 1953, James D. Watson, F. H. C. Crick, and M. H. F. Wilkins discovered the structure of the DNA molecule (see Figure 1-1). Since that time extensive studies on the nature of DNA action and its control of living systems at the biochemical level have led to new levels of understanding of the molecular basis of genetics, development and metabolism. The establishment of the nature, structure, and action of DNA has led to a revolution in biological thought unequaled in the present century, leading to new directions of thought about all aspects of biology.

Both the Mendelian and molecular views of biological heredity contribute to evolutionary theory by explaining reproduction. Our present understanding of reproductive processes may be summarized as follows.

1. Development of individual organisms is controlled by a series of hereditary regulators called *genes*. Each gene is equivalent to a portion of a DNA molecule and regulates the production of a single polypeptide chain through production of RNA (see Figure 1-5). In all living systems, the proteins, including those serving as major structural components, and the enzymes responsible for controlling activities are composed of a single polypeptide chain or a series of polypeptide chains formed individually in this fashion and subsequently aggregated. Essentially one gene forms one polypeptide chain.

2. Genes, in almost all organisms, are found in large units called *chromosomes;* chromosomes are usually included in the cell nucleus. The central axis of the chromosome appears to consist of a single, very long DNA molecule comprising hundreds of genes. Around the DNA skeleton are other complex macromolecules: proteins called *histones*, residual protein, and RNA manufactured by the DNA (see Figure 1-4); the number and type of chromosomes are usually constant for each species, as is the amount of DNA for each somatic cell. Some organisms, particularly bacteria and blue-green algae, lack a nucleus but apparently have a single genetic aggregate that functions as a single chromosome, composed of a

DNA double strand, located in the cell material but lacking a protein complex around it. Viruses are made up of a single protein-coated strand or a double helix of hereditary material and correspond to several free-living genes; in some viruses the hereditary material consists of ribonucleic acid (RNA) rather than DNA.

3. DNA is capable of coordinated and exact replication of itself (see Figure 2-5), giving rise to exact duplicates of the original DNA and consequently of the original genes. This process is the key to biological heredity, and explains the continuity of common features in all living systems and the conservative nature of biological reproduction.

The majority of organisms are cellular and contain a nucleus and discrete chromosomes; subsequent discussions will refer only to examples of this type.

All organisms that contain chromosomes are of two general types. In most species the chromosomes within most cells occur in pairs, so that each cell contains two sets of "homologous" (corresponding) chromosomes, each with homologous DNA double strands, and each, therefore, containing paired homologous genes. This kind of cell is called *diploid* because it contains twice $(2N)$ the basic number (N) of chromosomes. For example, all cells but the reproductive cells of the human body contain 46 chromosomes or 23 pairs, the usual diploid number for man. The total number of chromosomes is $2N$—that is, (2×23) or 46. In some organisms and in the reproductive cells *(gametes)* of higher plants and animals, the chromosomes are unpaired. This kind of cell contains one chromosome from each homologous pair or is *haploid* (N). The haploid cells of man, for example, contain 23 or N chromosomes. In most cases in diploid organisms with haploid gametes, exactly one-half the chromosomes, namely one from each pair of homologous chromosomes, are present in each haploid cell. Again, if man is used as an example, each sperm (male gamete) or egg (female gamete) contains N, or 23, chromosomes, one from each pair found in the diploid cells. It is also interesting to note that the amount of DNA present in a diploid cell of a particular species is twice that found in a haploid cell.

CELLULAR REPRODUCTION   In organisms other than viruses, reproduction is based on the origin of new cells by growth and division of other cells. In all multicellular organisms, whether they are haploid or diploid, new cells are being produced constantly by cell divisions to replace worn or injured cells and to provide for growth. The new cells contain exactly the same number and kinds of chromosomes present in the original cells. In addi-

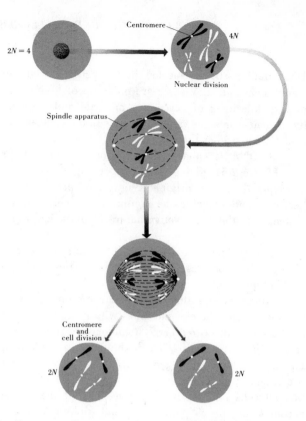

Figure 2-1 *Diagram of mitosis in an organism with the diploid number (2N) of chromosomes 4; changes are shown in the nucleus.*

tion, new individuals may be produced by this same type of cell division in organisms with *asexual reproduction*. In this mode of reproduction a cell or group of cells divides from the parent and becomes an independent individual. The new individual begins as a single cell, but by cell division it grows into an organism composed of thousands of cells. In many one-celled organisms, new individuals are produced simply by the growth and division of the parent (cell) into two new individuals (cells) containing the identical chromosomal complement of the parent cell. The numbers and kinds of chromosomes in individuals produced by asexual reproduction are exactly the same as those in the parent.

The process of cell division that produces new cells or individuals with replicas of the chromosomes of the parent is called *mitosis*. The significant features of mitosis are presented in Figure 2-1. (In the example, mitosis is followed in a diploid organism with two sets of chromosomes; that is, $2N = 4$.) During the periods between divisions, the cell is spoken of as being at the resting stage (interphase). As the cell begins to prepare for division, toward the end of the resting stage the chromosomes dupli-cate essentially exact replicas of themselves. The process of mitosis begins when the chromosomes condense into compact bodies. Each kind of chromosome is represented at this stage by a pair of duplicates attached by a small body called a *centromere*. In the accompanying diagram (Figure 2-1), the homologous chromosomes of each of the two sets are of the same size but of different shades, and replica chromosomes are of the same size and shade. Shortly after the condensed chromosomes appear, the nuclear membrane breaks down, the spindle apparatus appears, and the chromosomes line up along the equatorial plane. The centromeres now divide and the daughter chromosomes move to opposite poles, apparently assisted by the spindle apparatus. The process is completed by the division of the cell and the formation of new nuclear membranes, producing two new cells each with a nucleus containing $2N$ chromosomes identical to those in the parental cell. The process is summarized in Figure 2-3.

A second kind of specialized division of cells is characteristic of some stages in the life history of all organisms with sexual reproduction. This process is called *meiosis* (Figure 2-2) and provides for reduction of the parental diploid chromosome number $(2N)$ to the haploid number (N). Meiosis may occur in unicellular organisms, but in multicellular forms it occurs only in the reproductive organs *(gonads)*, all other cellular divisions in the body being mitotic. Thus, in human beings cells are produced by mitosis throughout the entire body, except in the gonads (male: *testis;* female: *ovary*) where meiosis produces the reproductive cells or gametes (male: *sperms;* female: *ova* or eggs).

The significant features of meiosis are indicated in Figure 2-2 for an organism with a diploid number $(2N)$ of four chromosomes; the number and arrangement follow those in the example for mitosis (Figure 2-1). Two cell divisions occur during meiosis. After the chromosomes are replicated during interphase, division begins with condensation, just as in mitosis. About the time the spindle apparatus appears and the nuclear membrane breaks down, the homologous chromosomes come to lie on opposite sides of the equatorial plane. The replicas are still attached to one another by the centromere, but unlike the situation in mitosis at this stage, the centromeres are not on the equatorial plane and do not divide. Subsequently the homologous units (two replicate chromosomes of each homologous pair) move toward the poles of the spindle apparatus. Nuclear membranes appear and cell division is completed from each homologous pair. This process constitutes the first meiotic division.

The second meiotic division begins with recondensation of the chromosomes and the breakdown of the nuclear membrane. No further replication of chromosomes occurs between the first and second divisions. The chromosomes now line up on the equatorial plane as in mitosis, the centromeres divide, and one replica of each chromosome moves to each pole. Nuclear membranes appear and the second meiotic division is completed, producing four cells, each containing a haploid number of chro-

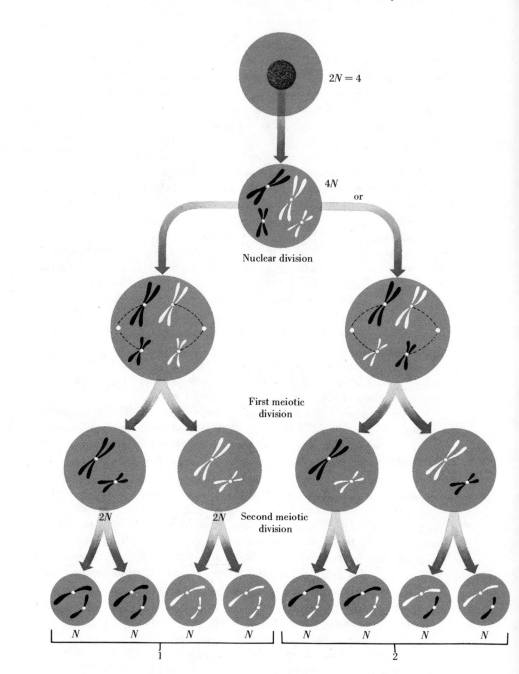

Figure 2-2 *Diagram of meiosis in an organism with the diploid number (2N) of chromosomes 4; haploid number 2; changes are shown for the nucleus.*

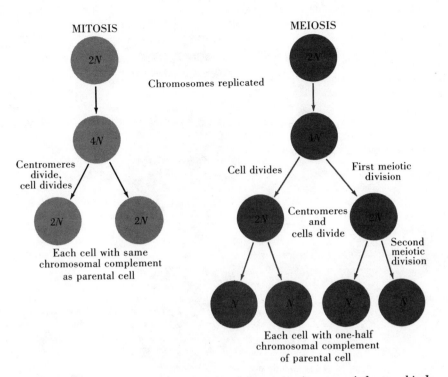

Figure 2-3   A summary comparison of the major features of the two kinds of cell division, mitosis and meiosis.

mosomes (N). The manner of separation is random; that is, a replica of either chromosome of a given homologous pair may go to either daughter cell during the first meiotic division (in Figure 2-2, alternate 1 or alternate 2). The new cells produced by the second meiotic division each contain one-half the chromosomes found in the parent diploid cell, or one chromosome from each homologous pair. The process is summarized and compared to mitosis in Figure 2-3.

Another aspect of meiosis, exceedingly important from the evolutionary viewpoint, is an event that occurs during the first meiotic division (Figure 2-4). Frequently during the process of separation of the chromosomes as the cell divides, the strands do not separate properly and material is exchanged between strands. The point of attachment and exchange is called a *chiasma* (plural, *chiasmata*) and is responsible for the genetic phenomenon of *crossing over*. Crossing over is an important source of genetic variability and will be discussed in some detail in Chapter 4. Figure 2-4 shows the formation of chiasmata and their effects on gametes.

If one chiasma occurs between different chromosomes, twice as many gamete types may be formed as when no chiasmata are present. (How

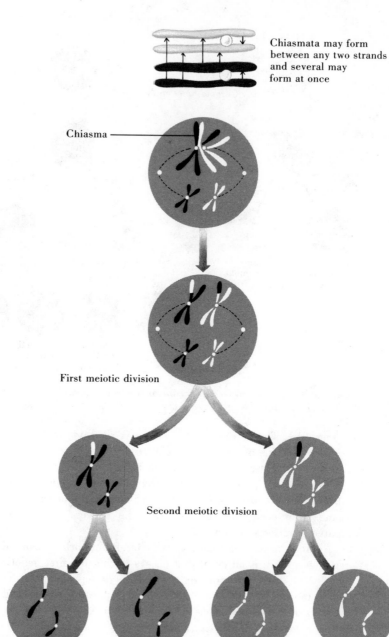

Chiasmata may form
between any two strands
and several may
form at once

Chiasma

First meiotic division

Second meiotic division

Figure 2-4 Diagram of chiasma formation during meiosis. The process causes portions of homologous chromosomes to be exchanged and increases the number of gamete types produced by a particular organism; only the nucleus is represented.

many kinds of gametes are possible when one chiasma has occurred during meiosis in an organism with $2N = 4$, such as in Figure 2-4?)

In sexual reproduction, new individuals are produced by the coming together of two gametes (usually a male and a female gamete) to form a diploid cell, or *zygote*. The zygote then develops into the new organism. In most sexually produced diploid organisms one-half their chromosomal and genetic material comes from each gamete, or one-half from each parent. Meiosis is responsible for this situation and insures that half of the heredity of the individual is carried by each gamete.

THE MOLECULAR BASIS OF REPRODUCTION   From the molecular point of view, the fundamental features of any living system are determined by a series of specific enzymes that regulate metabolism, development, and growth. It is because of the specific nature of enzymes, and their interactions and effects on other kinds of molecules, that each kind of organism exhibits a peculiar complement of physiologic, morphologic, and ecologic features that distinguishes it from other organisms. For this reason oak trees differ from lions and lions differ from tigers. As discussed in the preceding chapter, protein (including enzyme) synthesis is regulated by DNA. The differences between oaks and lions ultimately involve differences in the DNA molecules, in their arrangement, and in their expression through influences upon the characteristics and activities of the organism centered on protein and enzyme synthesis. Since DNA controls the specific features of each kind of living organism, it follows that the key to understanding biological reproduction is in the mechanism by which a specific DNA structure is reproduced and passed on to new cells or individuals, which will resemble their parents in all essentials. The general process of DNA replication is now well established. During the resting stage, described above under Cellular Reproduction, the DNA double helix strands become separated through the breaking of the weak chemical bond between complementary base pairs (Figure 2-5). Each strand acts to form a complementary strand by pulling complementary bases out of the surrounding cellular material. By the time cellular division is initiated, each strand has completely reduplicated itself, so that there are now two replicas of the original double-stranded helix of DNA. By this means, each daughter cell receives a duplicate of the DNA of the parent cell. This process is called the *semiconservative replication* of DNA, since each new double helix contains one new strand and one strand from the original cell. Because the order of the bases on the double strands of DNA determines the kinds of proteins and enzymes produced by the cell, it is apparent that cells produced by mitosis

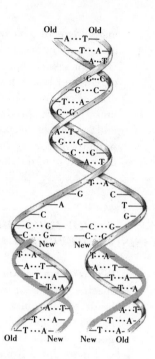

Figure 2-5   Diagrammatic representation of reproduction of DNA molecules: replication. The two strands making up the DNA double helix separate, and each induces formation of a complementary strand to produce two double helix molecules that are replicates of the original molecule. Each new molecule is made up of an old strand from the original helix and a new strand produced during the process of replication.

will contain essentially identical DNA complements and genes, and will produce essentially identical proteins and enzymes.

In meiosis the four haploid cells produced from one chromosomal replication (and one DNA replication) will each contain DNA double strands produced by exactly the same kind of semiconservative replication discussed above. The zygote, or new individual produced by the fusion of two haploid gametes, receives half its DNA complement from each parent, in the form of one homologous DNA double-stranded helix from each gamete. In some cases the amount of DNA present in diploid cells is not obtained equally from parental gametes. In many insects, fishes, and other vertebrates, males and females differ in the chromosomes (and amount of DNA) present. In human beings, for example, females usually have 23 similar pairs of chromosomes, one pair called the X chromosomes; males have 22 similar pairs and two other chromosomes (XY) that differ from one another in size and shape, but which pair up at cell division. Consequently, female gametes usually contain 23 chromosomes one of which is an X chromosome, but half the male gametes contain 23 chromosomes corresponding to the group in the female gametes (including an X), and the other half 22 plus the short Y chromosome. Strictly speaking, all normal human gametes contain 23 chromosomes, but the amount and kind of DNA and corresponding genes are less in those gametes containing a Y chromosome. New human beings arise from zygotes that are XX or XY, depending upon the chromosome complement of the father's gamete. The former (XX) are female, the latter (XY) male.

In many organisms one sex is characterized by an unpaired chromosome (XO), and the other sex has the same chromosome paired (XX); in other organisms the chromosomes in males are XX and females XY. In these one sex produces two kinds of gametes in terms of chromosome numbers and total DNA, the other only one kind.

These patterns form interesting deviations from the usual. In most sexually reproducing organisms, exactly one-half the chromosome complement, one-half the genes, and one-half the DNA come from each parent through the haploid gametes. Gametes produced by the diploid parents of these organisms receive exactly one-half the parental chromosomes and genetic material. Meiosis operates through the process of gamete formation to assure this result. For the development of concepts of evolution subsequently presented in this book, all examples assume, for simplicity, that one-half the chromosomes, genes, and DNA present in the diploid adults come from each parent through the haploid gametes.

GENES AND ALLELES   In previous sections it has been shown how chromosomes and DNA are replicated and inherited. Most genetic studies are concentrated on the individual segments of the DNA molecule, the genes, each capable of directing the formation of a particular polypeptide chain. The gene and its polypeptide product are apparently colinear, and the genes are arranged in a linear sequence on the chromosome. On paired homologous chromosomes of a diploid cell, each chromosome parallels the other in the arrangement of the genes, and two homologous genes — one from each chromosome — are present in each diploid cell. Very frequently a particular gene location is represented on different chromosomes by several slightly different base sequences that produce different polypeptides. These differences in polypeptides may be reflected in enzyme formation and in the visible expression of some characteristic. The different base sequences, or expressions of them, are called *alleles* of the gene. In tomatoes, for example, the gene for stem color may produce one or the other of two enzymes that control the presence of a purple pigment. One allele (base sequence) if present produces a purple stem, the other a green stem. The gene is called the stem-color gene (although it does not produce stem color by itself), and its two expressions are the purple-stem and green-stem alleles. Each diploid tomato plant cell contains two stem-color genes, one on each of a homologous pair of chromosomes. Tomato plant gametes always contain only one stem-color gene, since they contain only one chromosome from each homologous pair.

In any particular individual organism, any combination of alleles for a particular gene may be present. A tomato plant may have two purple-stem alleles or two green-stem alleles, or one of each. For simplification, we can symbolize these alleles as

$$a_1 = \text{purple stem,} \qquad a_2 = \text{green stem}$$

The possible combinations are

$$a_1a_1 \qquad a_2a_2 \qquad a_1a_2$$

When dealing with gene combinations the appearance of the organism is called its *phenotype;* thus the phenotype of a_1a_1 is a purple stem. The actual combination of genetic materials is called a *genotype;* thus the genotype of tomatoes with a green stem is a_2a_2. In many instances, when two different alleles for a particular gene are present, the phenotype appears identical to that of an individual having two identical alleles. The other allele is present and is passed on to the gametes but is masked out by what may be called the *dominant* allele. The hidden gene is called a *recessive* allele. In tomatoes, individuals with the genotype a_1a_2 have a purple stem (phenotype), indicating that a_1 (purple) is a dominant allele and a_2 (green) is recessive.

Of course, of utmost importance in considering heredity and the probabilities of gene combination is the production of gametes. Individuals of the genotypes a_1a_1 and a_2a_2 each produce only one kind of gamete, all a_1 and all a_2, respectively, and are called *homozygous.* Organisms with a genotype of a_1a_2 produce equal numbers of each type of gamete and are called *heterozygous.*

PROBABILITY AND GENETICS

Because of the constant and exact manner in which chromosomes and genes are replicated in reproduction, the theory of probability may be applied to the analysis of genetic and evolutionary events. In mitosis no range of possibility exists, for each daughter cell is an exact genetic duplicate of the parental cell. Probability theory is of great significance in understanding the genetics of sexual reproduction, however, because new individuals produced by this means receive, in the form of discrete units, exactly one-half their heredity from each parent.

Most readers are familiar with probability theory as applied to the toss of a coin: What is the probability of obtaining a head on any particular toss? Or in playing cards: What is the probability of cutting the deck to a king or to the ace of spades? A better example for our discussion is to reconsider the case of meiosis developed above (Figure 2-2). Let us regard the products of two lineages (1 and 2) as representing the gametes of two individuals, one (1) a male, the other (2) a female. Further, we will concern ourselves only with the single pair of large chromosomes, which can be either dark or light. If it is assumed that 1000 male gametes are present, it is obvious from the nature of meiosis that 500 will contain a dark chromosome and 500 will contain a light chromosome. Under these circumstances let us first consider the probability of sampling the gametes and obtaining one with a dark chromosome. It is clear that the chance of drawing a gamete with a light chromosome is just as likely as drawing a dark one; the events are equally probable. How do we express these probabilities?

The general formula for the probability, R, of a single event is

$$R = \frac{f}{f + u}$$

where f is the number of ways in which the selected or favorable event may occur, and u the number of ways in which some other outcome or unfavorable event may occur. R is usually given as a fraction or decimal; $f + u$ is always equal to the total number of events. To return to the example, the number of favorable events is 500 dark chromosomes, and the total number of events possible is 500 dark chromosomes plus 500 light chromosomes:

$$R = \frac{500}{1000}$$

$$R = \frac{5}{10}$$

$$R = 0.5$$

The probability of obtaining a gamete with a dark chromosome on any one draw is 0.5. (What is the probability of obtaining a gamete with a light chromosome?)

The probabilities for any single event range from 0 to 1.0, a probability of 1.0 representing certainty. The probability of drawing a gamete with two chromosomes from one of the individuals (1 or 2) indicated in the diagram is 1.0. A probability of 0 means that the event is impossible. The probability of obtaining a gamete with 7 chromosomes from the indicated individuals is 0. In any given situation, the sum of the probabilities of all possible events *always* equals 1.0. (With these facts in mind, what is the probability of drawing a gamete containing one large dark chromosome and one small light chromosome from the mixed gametes of individuals 1 and 2, if the total number of gametes present is 1000?)

Thus far, we have considered only the probability laws relating to single events; those principles applicable to the probabilities of two or more independent events happening simultaneously are more meaningful for problems of sexual reproduction. Again if we consider only the larger chromosomes in the gametes (1000 male gametes and 1000 female gametes), what is the probability of a zygote being formed by the coming together of any two gametes (one male and one female)? To answer this question, let us assume that fertilization has occurred at random to produce 1000 zygotes. What is the probability that a zygote will contain two light chromosomes? The probability of two or more independent events happening jointly is the product of the probability of one event times the probability of the other:

$$R = \frac{f_1}{f_1 + u_1} \times \frac{f_2}{f_2 + u_2}$$

$$R = R_1 \times R_2$$

In this example, the probability of obtaining a male gamete with a large light chromosome is 0.5 (R_1), and the probability of a female gamete with a large light chromosome is 0.5 (R_2). The probability of any zygote in the sample having two large light chromosomes is

$$R = 0.5 \times 0.5$$

$$R = 0.25$$

The probability of obtaining a zygote from a male gamete with a large dark chromosome and a female gamete with a large dark chromosome is similarly 0.25. Again, the probability of drawing a zygote with a large light chromosome and a large dark chromosome formed from a male gamete with a light, and a female gamete with a dark, chromosome is 0.25; the probability of the reverse situation, a zygote formed from a male gamete with a dark, and a female gamete with a light, chromosome, is also 0.25. In drawing zygotes from the pool of 1000 there are only these four possibilities, each with a probability of 0.25. The sum of the four probabilities equals 1.

In the example above, it is seen that there is only one possible way to produce a zygote with two light or two dark chromosomes; however, a zygote containing one light and one dark chromosome may be formed in two ways. If we are concerned only with the probability of getting a zygote of this latter type, a third probability principle is applicable.

When a particular independent event may occur in more than one way, its probability is the sum of the probabilities for each manner in which the event occurs:

$$R = R_1 + R_2$$

In the discussed example the probability for each of the two ways of obtaining a zygote with one light and one dark chromosome is 0.25. The probability of drawing such a zygote, without reference to the manner of origin, is

$$R = 0.25 + 0.25$$
$$R = 0.5$$

Although the manner of determining probabilities to explain the three principles has been detailed above, for our purposes it is not necessary to work out probabilities by listing all the possible arrangements in every case. A simple formula will serve to provide us with all the required information. Since a zygote is always formed by the joint occurrence of two independent events (the two gametes), the expansion of the binomial $(p + q)^2$ provides the probabilities directly.

$$(p + q)^2 = p^2 + 2pq + q^2 = 1$$

$p =$ frequency or probability that a zygote contains a light chromosome $(p = 1 - q)$

$q =$ frequency or probability that a zygote contains a dark chromosome $(q = 1 - p)$

The exponent 2 indicates that two independent events involving p and q are occurring simultaneously; $p + q = 1$, $(p + q)^2 = 1$.

In the expansion, each term indicates the probability of a particular combination of the two events:

$p^2 =$ frequency or probability of a zygote with two light chromosomes $(p \times p)$

$q^2 =$ frequency or probability of a zygote with two dark chromosomes $(q \times q)$

$2pq =$ frequency or probability of a zygote with one light and one dark chromosome (2 indicating that there are two ways to obtain this result)

This general formula may be applied to any situation involving zygote formation. For example, if the large chromosomes are ignored (Figure

2-3), what are the probabilities of drawing zygotes (produced by crossing 1 and 2) containing two small light chromosomes and two small dark chromosomes or a small light and a small dark chromosome?

The preceding section of this discussion indicates how the hereditary carriers, the chromosomes, are inherited. The actual regulators of heredity, the genes, are located in a linear sequence on the chromosomes. On the basis of the above discussion and your background in genetics, solve the following problems.

PROBLEMS

1.   A plant breeder carried out the following experiment involving garden peas. He crossed a strain homozygous for smooth-coated peas (in the pod) with another homozygous for wrinkled peas. All the offspring had pods full of smooth-coated peas. Since garden peas are usually self-fertilizing, he prevented self-fertilization and made random crosses among these individuals. What are the probabilities of finding each of the following in the next generation: phenotypes for wrinkled and smooth; genotypes for homozygous recessive, homozygous dominant, and heterozygous?

2.   A plant breeder crossed a homozygous red flowered primrose with a homozygous white-flowered primrose. All the offspring were pink flowered. One of the pink-flowered individuals was crossed back to the original red-flowered plant. What are the probabilities for each of the following in the next generation: phenotypes for red, white, and pink; homozygous and heterozygous genotypes?

3.   In many animals more than two alleles may be found for any particular gene. In rabbits, four alleles for the coat-color gene are present. One allele when homozygous produces a dark gray coat; the second a light gray coat; the third a Himalayan coat, white and black ears, nose, feet, and tail; and finally albino. The alleles show dominance, decreasing in the order given above. A Himalayan individual is crossed with an albino. If it is assumed that all genotypes are equally represented, what is the probability that the offspring will be albino? The same Himalayan is crossed with an individual with a dark gray coat; what is the probability that the offspring is albino?

POPULATION   Up to this point, we have emphasized genes,
GENETICS   the fundamental units of biological heredity.
Genes and biological heredity are conservative and ensure the continuity of efficient adaptations. Genes tend to be stable elements capable of exact duplication of themselves at the time of reproduction. They are important in that no evolutionary changes are possible unless a new genetic expression (an allele) has developed through faulty gene replication. Genes, however, are not the principal units of evolutionary change.

Genes do not occur as discrete, naked elements. With the exception of viruses, all living organisms consist of a complex mass of nongenetic material under the regulation of many interacting genes. To overgeneralize, it almost appears that all the complexities and adaptations found in living organisms serve simply to protect, nourish, and make possible the orderly reproduction of genes. The differences between diverse kinds of organisms — between a tiger and a leopard, or a beaver and a redwood tree — are not due to simple differences at the gene level. Each individual is the product of the interplay of his total genetic complement. The sum total of each individual's genotypic adaptation (as expressed through the phenotype) to a particular set of environmental conditions determines that individual's relative adaptive efficiency. Although genotypes are usually carried in the organic packages we call individuals, and although individuals may be affected by evolutionary forces, individual organisms are not primary units of evolution: neither individual genes nor individuals evolve.

As long ago as the early 1930s Ronald A. Fisher in Great Britain and Sewall Wright in the United States established the principle that the evolutionary forces act not to produce change in individual genes, individual gene combinations, or individuals, but to produce change in those groups of individuals called populations. A *biological population* may be defined at this point as all the individuals of the same species occurring in the same area at a particular time. In addition, a biological population, unlike individual genes and individuals, has a continuity through time, as reproduction adds new individuals to the population while death removes others. Insofar as evolutionary processes are concerned, a population consists of the pooled genes of all individuals within the group, expressed indirectly through various genotypes as individual phenotypes. The most significant source of evolutionary change is the impact of natural selection on this gene pool. The population forms the stage or setting for evolution and provides the matrix for the operation of the basic evolutionary forces. The population is the fundamental unit of evolution. In it, individual genes, genotypes, and phenotypes may be eliminated or reproduced and certain genotypic combinations may be favored over others, but the forces operate on the total gene pool. *Evolution,* at its simplest, is broadly defined as any change in the hereditary composition of a population.

A final genetic concept of utmost importance to an understanding of the elemental forces of evolution is the basic idea of population genetics. Each population is essentially a unit with a common body of genetic material. The fundamental discovery about these gene pools was made independently in 1908 by G. Hardy in Great Britain and W. Weinberg in Germany. These geneticists found that in the absence of factors which change gene frequencies, populations may have any proportions of different alleles, and that the relative frequencies of particular gene alleles and genotypes will remain constant generation after generation. We now know that the Hardy-Weinberg theorem is another expression of the hereditary conservation of DNA replication and genes. The resultant genetic stability of populations under these conditions is spoken of as *genetic equilibrium* or *genetic inertia*. Evolution occurs only when the equilibrium is upset or the inertia overcome.

An example of populational genetic analysis and the Hardy-Weinberg genetic equilibrium may be provided by study of an actual experiment. The experimenter isolated an artificial population of chickens in two large chicken yards. Two hundred chickens, equally divided between the two yards, formed the initial population. The genetic feature studied was one of the genes regulating plumage color in these animals. On the basis of other work it is known that two alleles of one gene produce black plumage (a_1a_1) or white plumage splashed with black (a_2a_2) in the homozygous condition, and heterozygous (a_1a_2) chickens have bluish-gray plumage. One hundred chickens in the experimental population had black plumage; the other 100 had splashed-white plumage. The chickens were allowed to breed at random so that every possible genotype would be produced.

A populational geneticist would ask the following questions: What are the frequencies of a particular allele in the population? What will be the frequencies of the genotypes and phenotypes in the next generation? These are simply questions of probability: What is the probability that any gamete will contain a particular allele? What is the probability that any zygote will contain any particular combination of two alleles?

In the experiment described above, the original population consisted of chickens homozygous for black plumage (a_1a_1) or splashed white (a_2a_2). Chickens with black plumage produce only one kind of gamete, whether sperms or eggs, with a genetic constitution of a_1. Splashed-white chickens similarly produce only gametes with a genetic constitution of a_2. Within the population of 200 individuals there is a gene pool of 400 plumage-color genes (200 black alleles and 200 splashed-white alleles), since each diploid individual contains two genes for plumage color. The gene frequency of a_1 is thus equal to 200/400 or 0.5, and the gene frequency of a_2 is also equal to 200/400 or 0.5. One-half the total gametes of the studied population are of each type.

From this information one may predict the genotypic and phenotypic frequencies in the next generation, a prediction verified by the results of the random crosses. With only two kinds of alleles there are only three ways in which the gametes may combine at random to form new chickens. Some a_1 gametes will combine with a_1 gametes. Other gametes will combine with a_2 gametes. Finally, some a_2 gametes will combine with other a_2 gametes. The probabilities of any two gametes coming together to form a zygote is the product of their frequencies. It is obvious that the same relationships apply here as in the examples in the preceding section on probability and can be developed by a binomial expansion: $(p + q)^2 = p^2 + 2pq + q^2 = 1$. In this case $p = 0.5$ and $q = 0.5$. Substituting in the formula gives us the probabilities for each genotype:

$$p = \text{frequency of } a_1$$
$$q = \text{frequency of } a_2$$

$$(0.5 + 0.5)^2 = \underset{\substack{a_1a_1 \\ \text{black}}}{0.25} + \underset{\substack{a_1a_2 \\ \text{blue}}}{0.5} + \underset{\substack{a_2a_2 \\ \text{splashed} \\ \text{white}}}{0.25} = 1$$

The significance of genetic equilibrium is demonstrated if we carry our experiment one more generation. In the first generation we had two genotypes and in the second generation three genotypes. What will be the situation if the population of second-generation individuals is allowed to breed at random with one another? We do not need to know the absolute number of individuals, because we already know that they are in the proportion of 25 percent black, 50 percent blue, and 25 percent splashed white. What are the gene frequencies of the second parental generation? Let us begin by determining the gene frequency of the allele a_1 in the second parental population. We know that for every 100 individuals in the population there are 200 genes. Twenty-five out of 100 individuals are of the genotype a_1a_1; that is, they contribute 50 a_1 genes to the total pool of 200. In addition, 50 out of 100 individuals have the genotype a_1a_2, and thus contribute 50 a_1 genes to the pool of 200. No other individuals have a_1 genes. Thus we see that 100 genes out of the total of 200 are a_1. The frequency of a_1 is 100/200 or 0.5. If $p = 0.5$, then q must also be 0.5, for $(q = 1 - p)$. The frequencies p and q both equal 0.5, just as they did in the previous generation. What will be the frequencies of the three possible genotypes in the next (third) generation? $p^2 = 0.25$, $2pq = 0.50$, $q^2 = 0.25$. These values added together equal 1.0. Is there something familiar here? Of course, because since the gene frequencies are the same, the genotype ratios are also the same in the new generation. In fact, unless conditions change, the gene frequencies and genotype frequencies will remain constant throughout all subsequent generations. This example may be extended into a fundamental rule: Once a population reaches genetic equilibrium $(p^2 + 2pq + q^2 = 1)$, the genetic frequencies will always be the same generation after generation, unless the equilibrium is upset.

Gene frequencies are rarely exactly equal in natural populations. A geneticist analyzed a group of tomato plants growing on an abandoned field as "escapes." He found that in 1000 plants selected at random, 510 had purple stems and 490 had green stems. What are the frequencies of the two alleles for stem color? What percentage of the population is heterozygous for the stem-color gene?

Or, as another example, a farmer started with a flock of 100 chickens of which 90 were splashed white and 10 were black. He allowed free inter-breeding for many years. What are the gene frequencies and genotype frequencies in this flock after 65 generations?

It is now apparent to biologists that the fundamental units of evolution are populations of genetically similar individuals, more or less isolated spatially from other similar populations of the same species. The pooled genes of the population tend to form a stable, unchanging equilibrium of genotypes that remains constant from generation to generation. Genetic equilibrium is an expression of the conservative nature of biological heredity, which favors the genetic *status quo* through exact replication of the DNA in cell division. Evolution can never occur unless this equilibrium is upset. The mode of evolution at the populational level is partially determined by the characteristics of population and by the relative frequencies of the raw materials of variation (the genes). The equilibrium may be upset if conditions change, either within the population through origin of additional variability or through external environmental changes. The elemental forces of evolution, mutation, natural selection, and genetic drift are radical elements that counter genetic equilibrium and bring about evolutionary change. If equilibrium is maintained, evolution is impossible. If the equilibrium is modified by any factor or factors, evolution occurs.

FURTHER READING

Cavalli-Sforza, L. L., and W. F. Bodmer, *The Genetics of Human Populations*. San Francisco: Freeman, 1971.

Levine, R. P., *Genetics*, 2d ed. New York: Holt, Rinehart and Winston, 1969.

Li, C. C., *Population Genetics*. Chicago: University of Chicago Press, 1955.

McKusik, V. A., *Mendelian Inheritance in Man*, 3d ed. Baltimore: Johns Hopkins Press, 1971.

Peters, J. A. (ed.), *Classic Papers in Genetics*. Englewood Cliffs, N. J.: Prentice-Hall, 1959.

Watson, J. D., *Molecular Biology of the Gene*, 2d ed. New York: Benjamin, 1970.

Whitehouse, H. L. K., *Towards an Understanding of the Mechanisms of Inheritance*, 3d ed. New York: St. Martin's Press, 1973.

Wilson, E. O., and W. H. Bossert, *A Primer of Population Biology*. Stamford, Conn.: Sinauer, 1971.

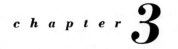

chapter **3**

The Elemental Forces of Evolution

Two fundamental patterns of evolution appear to be responsible for the changes that have produced the incredibly complicated labyrinth of diversity comprising the world of life. Evidence from the fossil record makes it clear that these two patterns have been repeated over and over during the history of life, while modern genetics provides a basis for understanding the forces producing the observed patterns. These patterns encompass diverse changes across the entire spectrum of organic diversity, from the evolution of characteristics of penicillin-resistant populations of the bacterium *Staphylococcus* to the magnificent radiations of the reptiles and flowering plants. It is important to note that the driving evolutionary forces at the base of such diversity remain the same, but that different expressions are obtained under different circumstances.

51

**BASIC
PATTERNS OF EVOLUTION**
Evolution at its simplest involves relatively minor changes in the gene pool of a particular population, from one generation to the next, with corresponding modification in genotypic frequencies and the range of phenotypic variation. No new populations result from the change, but the descendent population is not genetically identical with its predecessor. Evolution of modified gene pools from preexisting ones is called *sequential evolution*. Paleontologists working with short-time series, geneticists studying laboratory populations, and field biologists investigating isolated natural populations repeatedly verify the reality of sequential evolution within the history of a single population.

An example of sequential evolutionary change is provided by a combined genetic and field study of the scarlet tiger moth (*Panaxia dominula*). During a period of 23 years (1939 through 1961) a group of British geneticists studied the fluctuations of gene frequencies in a single population of this form. The species is ideally suited for such study because none of the adults of the previous generation survive the winter. Figure 3-1 indicates the fluctuations of gene frequencies for one pair of alleles in this population. In one expression of the gene, the black anterior wing of the moth has numerous light spots (a_1a_1); the second allele produces a wing with two light spots (a_2a_2). In heterozygous individuals there are several light spots in the wing. The fluctuations in gene frequency are given for normal and two-spot alleles. Note the significant changes between some generations (1940 and 1941) and the slight changes between others (1947 and 1948). No sign of genetic equilibrium exists here. Indeed, once we realize that in every population thousands of genes are present and subject to change, it becomes apparent that equilibrium is exceptional and sequential evolution the rule. All available evidence indicates that almost all populations undergo change of this sequential type every generation.

Sequential evolution may produce rather random fluctuations over long periods of time, with relatively little difference between the genotypic combinations and frequencies at the beginning and end of the studied period. It may also result in gradual shifts in gene combinations so that the descendent population is markedly different from its original ancestor. Sequential evolution, by itself, never produces new populations from old; it produces only temporal changes in a population continuum. In one sense, sequential evolution reflects the conservative nature of biological inheritance.

The second major pattern is one of *divergent evolution*, that is, the origin of new populations or organisms from old ones. It usually results from forces operating over a longer period of time than those responsible for sequential changes. Divergent evolution is most familiar to scientists working with the fossil record, but the diversity of the organic world at

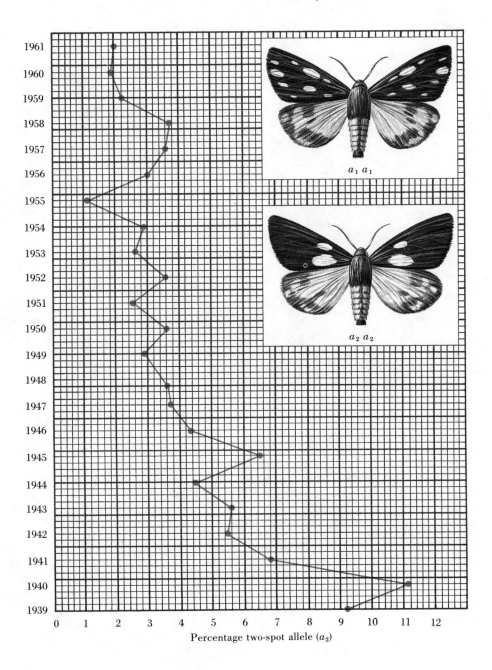

Figure 3-1 Sequential evolution; fluctuations in gene frequency in a population of scarlet tiger moths during a 23-year period. The homozygous genotypes, many-spot (a_1a_1) and two-spot (a_2a_2), are illustrated.

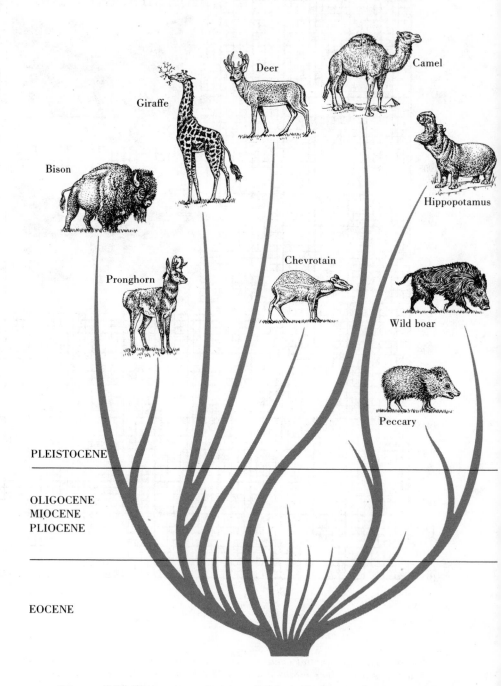

Figure 3-2 *Divergent evolution: adaptive radiation in the even-toed hoofed mammals.*

the present time is the result of divergence after divergence over the course of three billion years. Paleontologists see divergent evolution documented in the rocks where new populations develop as fragments of old populations, and divergence compounded on divergence radiates over the millions of years of records in a vast if not overwhelming array of organic diversity. Divergent evolution is responsible for the myriad modifications superimposed on the basic framework of living material to produce the spectrum of organisms from tuberculosis bacterium to towering redwood to humble flea. An exciting example of evolutionary divergence is provided by the radiation of the even-toed hoofed mammals, illustrated in Figure 3-2. The basic ancestor of this line was also ancestral to the odd-toed hoofed mammals and the carnivores.

The difference between sequential and divergent evolution as described in the above paragraph perhaps overemphasizes the distinction between them. Sequential evolution is not known to be exclusively characteristic of any population, since all populations seem to fragment or undergo divergent evolution if they are studied over a long enough period. Sequential evolution continues in the diverging new populations. However, although the elemental evolutionary forces responsible for sequential change also contribute to a great degree to divergent evolution, additional factors must cooperate to produce the latter. The two processes are intimately related but they are not identical. Sequential evolution and divergent evolution are graphically compared in Figure 3-3 to emphasize the obvious differences within the qualifications mentioned above.

The reality of the differences between sequential and divergent change must not be overlooked in discussing evolution. Only a few years ago, many workers thought that divergence could not possibly be explained

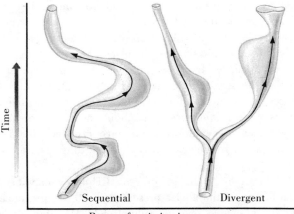

Figure 3-3   Comparison of sequential and divergent evolution.

in terms of the elemental evolutionary forces, while many others failed to see that divergent evolution involves more factors than those responsible for sequential change. Even today the relationship between the two processes, although clear in most aspects, remains the subject of some controversy. (The argument developed in Chapters 8 and 9 is designed to clarify and interrelate the distinctive features of both patterns of evolution.)

The remainder of this book is divided into two sections, in recognition of the two basic patterns of evolution, sequential and divergent. The first section treats the elemental forces of evolution common to both processes and the manner in which these forces combine to bring about sequential change. The final section of the book deals with the way in which the elemental forces are modified by the impact of additional factors leading to evolutionary divergence.

THE ELEMENTAL FORCES AND MICROEVOLUTION

The processes of sequential and divergent evolution are driven by the same elemental evolutionary forces, although additional factors contribute to divergence. These forces guide the course of populational evolution, control all sequential change, and provide the basic impetus for evolution in divergent new populations. We have already seen that the mechanism of biological heredity operates to produce genetic equilibrium. If all factors remain constant, the gene frequencies never fluctuate from generation to generation, but maintain this equilibrium. In addition, we have seen that equilibrium in naturally occurring populations is rare, if it ever occurs at all. In the case of the scarlet tiger moth (Figure 3-1), for example, evolutionary change occurred in each generation, sometimes only slightly modifying the gene pool, at other times markedly modifying it. What are the forces responsible for these sequential changes? How do they operate to produce evolution?

Evolution at its simplest involves changes in the frequencies of a single pair of alleles from one generation to the next. If we refer to the example of the scarlet tiger moth (Figure 3-1), we may briefly consider the causes, or elemental forces of evolution, responsible for the shifting pattern of gene frequencies throughout the 23 years. Any one of three primary forces of evolution could have produced the recorded changes seen between the frequencies of each succeeding generation.

1. *Mutation.* A change in gene frequency may be produced by an increase in any of the alleles present or by the appearance of new alleles, through spontaneous change (mutation). Mutation as an evolutionary force is the ultimate source of new alleles and new gene combinations. Variation provides the hereditary materials to be molded by the impact

of the other two forces. In most instances other factors contribute to the effect of variation by modifying and amplifying the effect of spontaneous gene mutation.

2. *Natural selection.* Environmental factors may operate to favor the differential reproduction of certain alleles or gene combinations over others present in the population. The impact of the total environment on the reproduction of gene combinations is the force of *natural selection.* The effects of natural selection change as the environment changes, so that slightly different environmental conditions in each generation favor slightly different gene combinations. Natural selection molds the genetic variation present in a population, but it cannot directly produce new genes or gene combinations.

3. *Genetic drift.* In many small populations completely random fluctuations in the frequencies of certain alleles or gene combinations may occur, even under constant environmental conditions. These fluctuations constitute *genetic drift,* which also makes its impact felt through random effects on the genetic variation already present in the population.

Mutation of two-spot (a_2) to normal (a_1) alleles might have been the mechanism that produced the change between the 1940 and 1941 generations of the scarlet tiger moth. Alternatively, a slight change in the environment may have enhanced the survival and reproduction of more normal alleles between the two generations, or, in other words, natural selection favored the normal over the two-spot allele. Finally, the shift might be due to random effects in this small population, or genetic drift, from two-spot to normal alleles. Unfortunately, we do not know exactly which force or forces produced the observed changes. Actually, all three forces probably are involved to some degree over the 23-year period, with the change between generations at any point being the result of interaction between all three forces rather than the effect of a single force.

Evolution resulting from interaction of variation, natural selection, and genetic drift to produce relatively small population changes is frequently called *microevolution.* Sequential evolution is always the product of the microevolutionary process; divergent evolution at its simplest is also microevolutionary. The rest of this part of the book deals with the process of sequential evolution through microevolution. The contributions of the elemental forces of variation, selection, and drift will be treated in detail with a discussion of their interactions in this basic process.

FURTHER READING

Adelberg, E. A. (ed.), *Papers on Bacterial Genetics.* Boston: Little, Brown, 1966.
Avers, C. J., *Evolution.* New York: Harper & Row, 1974.

Ehrlich, P. R., R. W. Holm, and P. H. Raven (eds.), *Papers on Evolution.* Boston: Little, Brown, 1969.

Ford, E. B., *Moths.* London: Collins, 1955.

Grant, V., *The Origin of Adaptations.* New York: Columbia University Press, 1963.

Lerner, I. M., *Heredity, Evolution and Society.* San Francisco: Freeman, 1968.

Moore, J. A., *Ideas in Modern Biology.* New York: Natural History Press, 1965.

chapter 4

The
Sources of
Variation

"No two individuals are exactly alike" and "All organisms exhibit variation" are biological clichés expressing the most obvious feature of living organisms. We need go no farther than any classroom to appreciate the reality of variation. Look around a class at the characteristics of the human individual in the sample. Almost any feature will demonstrate the point: hair color, eye color, skin pigmentation, or singing voice — each exhibits a spectrum of variation. No two individuals are precisely alike, because they are a combination of variants of a great many attributes. Two individuals may agree in having blue eyes, but differ from each other in almost every other respect. The overwhelming number of possible combinations of characteristics produces biological individuality.

Individuals within the same species may differ as a consequence of either heredity or environment. However, only the differences inherited by the next generation are of importance to evolution. The offspring of a man who has lost a finger in an accident are not born lacking the same finger, but the variation exhibited by human beings in hair color, eye color, skin pigmentation, and singing voice are all under the control of genes within the human population pool. The combinations of these genes with thousands of others result in the formation of distinctly individual human beings. Hereditary differences under the control of the genetic makeup of the individual are the raw materials of evolutionary change. But where did the genetic variation come from in the first place? Why are there blue eyes, green eyes, gray eyes, brown eyes, and black eyes? What are the sources of hereditary variation?

These are among the most trenchant of evolutionary questions, since we know that evolution cannot take place without variation in hereditary characteristics. Darwin recognized the central position of hereditary variation as a factor in evolutionary change, but it is only with the discoveries of modern genetics that the bases of heredity and the sources of inherited variation have been elucidated. Without a knowledge of these features our current concept of evolution would remain at a far less sophisticated level.

Thanks to the brilliant work of geneticists from a wide variety of fields and eras, biologists now recognize several sources of genetic variability.

A. MUTATION

   1. Gene mutation
   2. Chromosomal mutation

B. RECOMBINATION

   3. Heterozygosis (cross between two kinds of homozygous parents)
   4. Random assortment of the genetic materials
   5. Crossovers (genetic exchange between chromosomes)

C. GENE FLOW OR IMMIGRATION

Mutations are regarded as the ultimate source of new and different genetic material appearing in a population. Recombination is responsible for spreading the mutants through the population and for developing new combinations of genetic materials with materials from old genotypes. The following discussion emphasizes the points of significance to evolution.

GENE MUTATION In the early days of genetic study, a small fruit fly (*Drosophila melanogaster*) was selected as an excellent laboratory animal for hereditary experiments. In one of the laboratory colonies of normal red-eyed fruit flies, there appeared in 1909 a single white-eyed male. No white-eyed fruit flies had ever been reported until the appearance of this odd individual. When this male bred with red-eyed females in the laboratory, it was discovered that the white eye is due to an allele recessive to the red-eye allele. This spontaneous appearance of a new gene expression is called a *gene mutation.* The mutant individual is always descended from normal parents of a pure-breeding line and its peculiarity is passed on to the next generation. Since this appearance of a white-eye allele, about 15,000 more spontaneous white-eyed mutant individuals have been found among nearly 60 million fruit flies examined in the course of genetic studies. Spontaneous changes are known to have occurred in a wide variety of genes, including those controlling wing size and those affecting bristle length or eye structure. Some mutations have occurred several times, others only once, but each has increased the genetic variability of *Drosophila.*

Gene mutations are now known for every kind of plant or animal that has been subject to genetic study — including corn, snapdragons, coffee, cacao, mice, molds, bacteria, and men — and they seem to be a universal fact of life. Many of the differences between individuals of the same population are ultimately the result of gene mutation and the subsequent establishment of genetic equilibrium between the original allele and the mutant. The other principal features of gene mutation are summarized below.

1. Genes are relatively stable. In *Drosophila,* about one gene mutates for every 20 gametes produced. Since each gamete contains about 20,000 genes, this is a low rate of change, although the figure is probably higher because many slight mutations pass unnoticed.

2. Each gene has a characteristic mutation rate. Mutation rates for a number of genes in corn *(Zea)* are summarized in Table 4-1.

3. Some gene mutations result from a loss of genetic material; many, however, involve a change in composition rather than a loss. Such genes are known to mutate from a mutant allele back to the original expression (reverse mutation). In *Drosophila* the forked-bristle (a_2) allele is a known mutant of normal bristle (a_1); however, occasionally the reverse mutation from forked bristle to normal bristle occurs:

$$(a_1 \xrightleftharpoons{\text{mutates}} a_2)$$

4. More than one kind of mutation is possible for any particular gene. Some genes mutate to several different alleles, the basis for the presence of several alleles for one gene (multiple alleles).

Table 4-1   Mutation in Corn (Zea)

Genes	Mutation Rate (per million gametes)
Seed color	492
Seed-color inhibitor	106
Purple seed color	11
Sugary seed	2.4
Yellow seed	2.2
Shrunken seed	1.2
Waxy seed	0

5. Most of the obvious mutations are deleterious to the organism, but most mutations are not obvious. The significance of the first portion of this statement is more apparent than real and indicates the emphasis on abnormalities and striking mutations as a method of genetic analysis. In *Drosophila*, approximately 500 genes are known through discovery of mutations. Since the chromosomes of this animal are estimated to contain 20,000 genes and since we now realize that many mutations involve subtle chemical and physiological changes not expressed as major visible modifications, it seems likely that many mutations are not deleterious. From the evolutionary point of view, mutation is a requisite for change; in this sense mutations are advantageous, not deleterious, to long-term populational survival.

6. Rates of mutation may be modified experimentally. The type of mutation has no relation to the nature of the agent affecting its rate; rather, the mutants will be of the same types however they may be produced. Agents known to produce such changes in mutation rates of genes in exposed chromosomes include x-rays, cosmic rays, ultraviolet rays, gamma rays, temperature, and a variety of chemicals.

7. Mutation is the result of a slight change in the chemical structure of the segments of the DNA molecules that are the genes; differences in DNA structure are translated through production of different enzyme systems, and the new or mutant DNA can be replicated and inherited by succeeding generations.

Succinctly stated, a gene mutation is any change in the chemical organization of the gene that is replicated and passed on to succeeding generations. Mutation is random in the sense that the nature of the environmental stimuli that activate the chemical change does not influence the locus or direction of mutation. A rise in temperature will increase mutation rates in *Drosophila*, but mutations occur in a great many genes and not necessarily in any that control temperature tolerance or adaptation in the fruit fly. In another sense, however, mutation is nonrandom, since under a given set of circumstances mutation rates and mutant genes appear generation after generation in a constant pattern. We also know that certain genes mutate to certain alleles under the regulation of other genes. Mutation, although at one time thought to be rare, occurs as slight and nonobvious changes at a significant rate. Many aspects of the mutational process still need investigation and in them lie opportunities for future elucidation of genetic and evolutionary problems.

Gene mutations are the ultimate source of hereditary variation. By themselves their impact upon populational variability may be slight. Gene mutants, however, usually have amplifications of great significance because of their interactions with other genes besides their homologous alleles. The indirect effect of these interactions is frequently of greater significance in the development of a variety of genotypes and phenotypes, and in their impact on the course of evolution, than is the direct effect of the mutant allele.

How genes interact and contribute to variation is seen in simplest form when we analyze the phenotypic results of the interactions of two genes (each with two alleles). A diagram of plumage inheritance in parakeets (Figure 4-1) shows the significance of interaction in increasing the number of phenotypes. The two sets of genes $(a_1a_2)\,(b_1b_2)$ interact to produce four plumage colors, depending upon the gene combinations present. The combination a_1-b_1- will be green. The combinations $a_2a_2b_1b_1$ and $a_2a_2b_1b_2$ will be yellow. The combination $a_1a_1b_2b_2$ will be blue, but so will $a_1a_2b_2b_2$; $a_2a_2b_2b_2$ will be white. This type of interaction is very common.

Even more complex, but based on the same pattern of gene interaction, are characteristics regulated by several different genes, each contributing only a small effect to the total phenotype. These groups of genes are called *polygenes*. An example of polygenes is provided by the factors regulating skin pigmentation in man. The actual number of genes controlling pigmentation is thought to be four (possibly more), with at least two alleles for each gene. The range of possible genotypes is 81, each with a slightly different degree of pigmentation in phenotypic expression.

Another source of variation is provided by multiple alleles. The regulators of human blood types are an example. Only a single gene is involved, but it occurs as three distinct alleles, a_1, a_2, and a_3. In addition, although both a_1 and a_2 are dominant to a_3, neither is dominant to the other. The possible phenotypes and genotypes for human blood types are summarized in Table 4-2. In the case of some genes, as many as eleven alleles are known with 66 genotypes.

Another feature of the genes that contributes to variation is the phenomenon of *pleiotropism*. Many genes are now known to affect more than one characteristic, and such genes are called pleiotropic. The allele that produces vestigial wings in *Drosophila* also causes a modification in the balance organs, certain bristles, and sperm and egg production, among

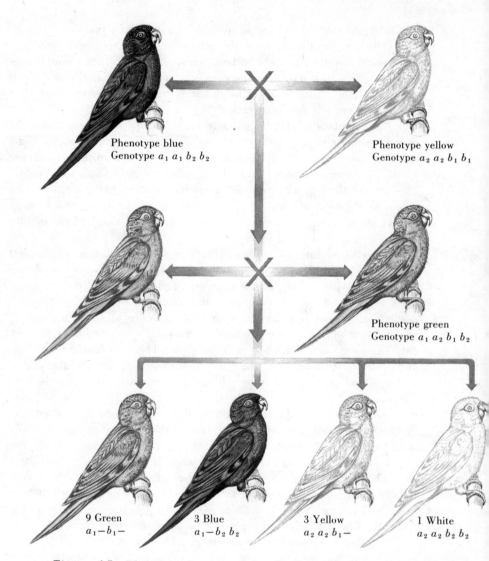

Phenotype blue
Genotype $a_1 a_1 b_2 b_2$

Phenotype yellow
Genotype $a_2 a_2 b_1 b_1$

Phenotype green
Genotype $a_1 a_2 b_1 b_2$

9 Green
$a_1{-}b_1{-}$

3 Blue
$a_1{-}b_2 b_2$

3 Yellow
$a_2 a_2 b_1{-}$

1 White
$a_2 a_2 b_2 b_2$

Figure 4-1 *Plumage inheritance in parakeets, under control of two independent genes.*

Table 4-2 **Human Blood Types**

Group	Genotype
AB	$a_1 a_2$
B	$a_2 a_2$ or $a_2 a_3$
A	$a_1 a_1$ or $a_1 a_3$
O	$a_3 a_3$

other characters. It seems likely that most genes have multiple effects, just as it is certain that many genes are involved in the development of any characteristic of an organism.

A final indication of the interdependence of genes lies in the discovery that the location of the gene in relation to other genes determines interactions and characters. In some of the chromosomal mutations discussed below, gene positions are changed and new characteristics appear although no gene mutation has taken place. The new characteristics are due to interactions between genes that are now positioned in a new relationship to one another.

All these comments on gene activity emphasize one important conclusion. An individual is the product of interactions involving his total genotype, and every gene plays a part in the process of development. A mutation that by itself is of little significance may contribute mightily to variation through its effect on gene action and interaction.

CHROMOSOME MUTATION   Chromosome mutations differ only in degree from gene mutations. If the chromosomes undergo spontaneous reorganization or modification, the effect is usually more pronounced than if only a single gene mutates. Chromosomal mutations are inherited once they occur and are of a considerable variety.

   A. Changes in number of chromosomes
     1. Loss, or gain, of a part of the chromosomal set: *aneuploidy*
     2. Loss of an entire set of chromosomes: *haploidy*
     3. Addition of one or more sets of chromosomes: *polyploidy*
   B. Structural changes in chromosomes
     1. Changes in number of genes
       a. Loss: deletion
       b. Addition: duplication
     2. Changes in gene arrangement
       a. Exchange of parts between chromosomes of different pairs: translocation
       b. Rotation of a group of genes 180° within one chromosome: inversion

Any of these major changes contributes to variability by changing the pattern of gene interaction, magnification of the role of position effect being particularly significant. Although chromosomal mutations are now well known and are sometimes striking in their phenotypic effects, for all practical purposes they may be treated in the same way as gene mutations in discussing microevolution (Figure 4-2).

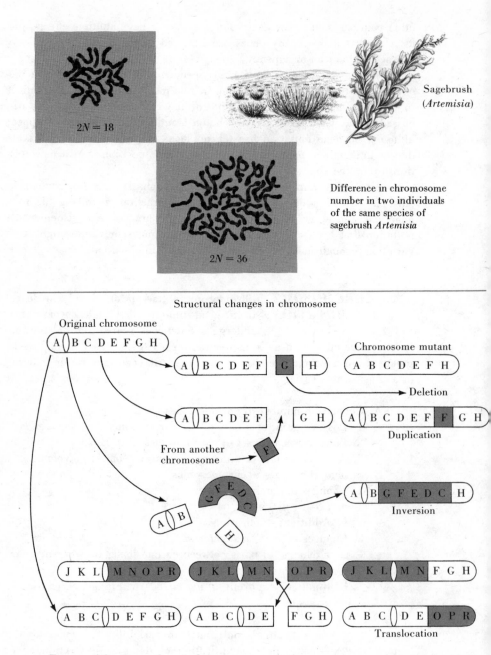

Sagebrush
(*Artemisia*)

Difference in chromosome
number in two individuals
of the same species of
sagebrush *Artemisia*

Structural changes in chromosome

Figure 4-2   Examples of chromosome mutations.

RECOMBINATION   In view of the nature of gene action and the great significance of cooperation between the genes, the process of mixing or recombining the available genes into a variety of genotypes rivals mutation as the primary source of variation. The full import of recombination as a source of genetic variation and as an equal partner with mutation in the manufacture of the materials of evolution is clearly recognized by biologists. Recombination—that is, new genotypes from already existing genes—is of several kinds: (1) the production of gene combinations containing in the same individual two different alleles of the same gene, or the production of heterozygous individuals; (2) the random mixing of chromosomes from two parents to produce a new individual; (3) the mixing of a particular allele with a series of genes not previously associated with it, by an exchange between chromosomal pairs during meiosis, called *crossing over,* to produce new gene combinations. The significance of these recombination processes may be shown by comparing the possibility for variation in both an asexually and a sexually reproducing population.

If we assume two populations each homozygous in every gene but one reproducing asexually and the other sexually, the importance of recombination and also of sex will be seen. First let us consider heterozygosity. A single mutant $(a_1 \rightarrow a_2)$ for one gene occurs in each population. In the asexual population, all the offspring of the single individual in which the mutation occurred will be heterozygous (a_1a_2), whereas all other members of the group will remain homozygous (a_1a_1), as will their offspring. Without heterozygosity, no variation would be present. In the sexual group, three genotypes will develop, since once the new allele is established it will combine at random in gametes. The genotypes are a_1a_1, a_1a_2, a_2a_2. In other words, from one gene mutation the latter population is 200 percent more variable in genotypes. Multiply this instance by 10 genes or 100 genes and the amount of potential variation produced by heterozygosity in a sexual population becomes fantastic, far outstripping the impact of the mutant alleles.

If we continue along this line, it will be immediately apparent that crossing over during meiosis does not occur in asexual reproduction and that the second contribution of recombination can operate only in sexually reproducing organisms. Mutation and restricted heterozygosity are the sole sources of variation for asexual reproducers. From a variational and evolutionary viewpoint, crossing over is an added dividend to the process of sexual reproduction. Sex is not essential for the reproduction of new individuals; it is a luxury. It is a luxury, however, that ensures greater potential for evolution and flexibility of adaptation than is ever possible for asexual forms.

Crossing over is produced during meiosis by chiasmata (see Figure

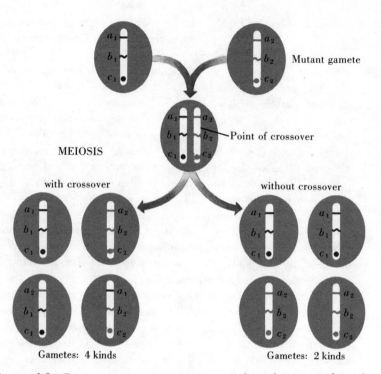

Figure 4-3 *Diagrammatic representation of the relative number of gene combinations possible as the result of one recombination crossover.*

2-4). It results in an exchange of genetic material between homologous chromosomes. In a more detailed example (Figure 4-3), imagine a population in which all individuals are of the following genotypes: a_1a_2, b_1b_2, and c_1c_2 for three genes on the same chromosomes. If no crossing over ever occurred, all gametes would be $a_1b_1c_1$ or $a_2b_2c_2$. However, since chiasmata are formed at almost every meiosis, new gametes and gene combinations often are produced. Crossover thus speeds the mixing of new mutant alleles with the genes already present. In this example four kinds of gametes appear: $a_1b_1c_1$ and $a_2b_2c_2$, plus two new ones, $a_2b_1c_1$ and $a_1b_2c_2$ and crossing over increases the number of possible genotypes from three to ten. If all possible crossovers occurred, six new gametes (the two already mentioned and $a_1b_1c_2$, $a_2b_2c_1$, $a_1b_2c_1$, and $a_2b_1c_2$) and a total of 36 genotypes (the three original plus 33 new ones) could appear. A more complicated, but also more realistic, example is found in the contribution of crossovers to human variation. As you will recall, there are 23 pairs of chromosomes in human beings, each with several thousand genes. Chiasmata may form at almost any point on these chromosomes, and a moment's thought will provide an appreciation of the utterly overwhelming potential for new gene combinations, and in turn for new kinds of individuals, provided by crossover. Since each point of chiasmata formation doubles the number of kinds of gametes, the gamut of human variation visible around us merely attests to the efficacy of recombination.

The significance of recombination to evolution cannot be overestimated. A single mutational change may be lost or passed on without great impact on a population, but if its effect is modified and enhanced by recombination, an unending contribution to variation is begun. Variation is the raw material for evolutionary change; recombination is its principal source. Mutation alone has relatively little effect on variation without the pervasive impact of recombination.

GENE FLOW OR IMMIGRATION   As long as a group of organisms exists as an isolated population, mutation and recombination remain the sole sources of genetic variation. Because natural populations rarely remain completely isolated from spatially adjacent populations of the same species over long periods, immigration of individuals (or dispersion of pollen, eggs, larvae, or seeds) from one population to another may produce changes in gene frequencies or introduce new alleles or recombinants not previously present. These effects parallel the impact of mutation rates if immigration continues or if new mutations or recombinations are added to the gene pool for the first time. The process of one population contributing genetic material to another is called *gene flow*, and in population genetics the rate of flow is measured by *immigration pressure* (m). This value will equal the proportion of alleles of a particular kind added to the population by the immigrant individuals. Reciprocal exchanges between populations frequently occur, and gene flow may lead to the amalgamation of the two populations into one.

THE MOLECULAR BASIS OF EVOLUTION   Gene mutations, although often expressed in visible characteristics or physiologic activities of the organism, are ultimately caused by rearrangement, addition, or deletion of specific nitrogenous bases in the DNA. Changes in the order of the bases will result in modifications in the polypeptide chain produced under the control of the gene, and these modifications will result in production of a modified enzyme (or other protein) or suppression of some enzyme (or

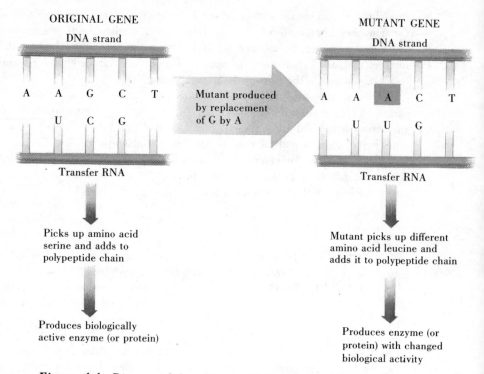

ORIGINAL GENE
DNA strand

A   A   G   C   T

U   C   G

Transfer RNA

Mutant produced
by replacement
of G by A

MUTANT GENE
DNA strand

A   A   A   C   T

U   U   G

Transfer RNA

Picks up amino acid
serine and adds to
polypeptide chain

Mutant picks up different
amino acid leucine and
adds it to polypeptide chain

Produces biologically
active enzyme (or protein)

Produces enzyme (or
protein) with changed
biological activity

Figure 4-4   *Diagram of the molecular (DNA) basis of gene mutation.*

protein) feature. The modified (mutant) enzyme may ultimately express itself by affecting a change in the physiology or development of the organism. A simple example of DNA mutation is provided in Figure 4-4.

Since a change in one of the bases may, theoretically, occur at any point along the length of one strand of the DNA, an incredible number of mutations are possible. Many mutations involve the addition or deletion of one to many bases along the DNA strand (Figure 4-5). Because the amino acids being added to the developing polypeptide chain are picked up by groups of three bases on the transfer RNA (Figures 1-3 and 1-4), a change in one base may effect which amino acid is added to the chain. In a gene composed of 1,500 base pairs, for example, a minimum of 1,500 single changes is possible, but these changes affect the sequence of only 500 amino acids in the polypeptide chain. Nevertheless, the number of slightly different mutants possible for such a gene is extremely high. Certainly, the possibilities for mutation along the extraordinarily long DNA molecules that form the axis of chromosomes provide more than adequate basis for variation and, therefore, for the raw materials of future evolutionary change.

VARIATION The simplest possible change in the genetics
AND of a homozygous population, all individuals
EVOLUTION of which have exactly the same genotype,
would be the production of a single mutant
allele $a_1 \rightarrow a_2$. If the mutation constituted a slight change of a not too
deleterious kind, or even if it were widely advantageous, its effect on
the population would be small. In fact, if the mutation occurred only once,
the population in the next generation would be at equilibrium, and no
change would subsequently occur unless conditions changed. Imagine, for
example, that the original population contained only 50 individuals (100
a_1 genes). After one a_1 mutates to a_2, the gene frequencies (p for a_1, q for
a_2) are $p = 0.99$, $q = 0.01$; the genotypes of the next generation are $p^2 =$
0.9801, $2pq = 0.0198$; $q^2 = 0.0001$. Forever afterward, if conditions
remain constant, the equilibrium remains the same, again illustrating the
unimportance of a single mutation.

As already indicated, however, gene and chromosomal mutations
usually occur at a regular rate each generation.

In corn, the mutation rate (μ) for a number of alleles has already
been indicated (Table 4-1). If a population initially homozygous for the
seed-color gene is studied, it is obvious that the gene frequencies of the
population will change from one generation to the next as follows.

$$\Delta p = -\mu p$$

Δp is the rate of change in gene frequency, and μ is the mutation rate
$a_1 \rightarrow a_2$. In the first generation the change may be computed

$$\Delta p = -(0.492 \times 10^{-4})\ 1.0$$
$$\Delta p = -0.0000492$$

The new gene frequencies will be

$$p = 1.0 - 0.0000492 = 0.9999508$$
$$q = 0.0000492$$

If the mutation rate is unopposed, equilibrium will be reached when the
population becomes homozygous for a_2. If the frequency for a_1 in any
generation is represented by p_o, the frequency for a_1 any number of generations (n) later is

$$P_n = P_o\ (1 - \mu)^n$$

Gene flow can be evaluated similarly by substituting the value mq_1

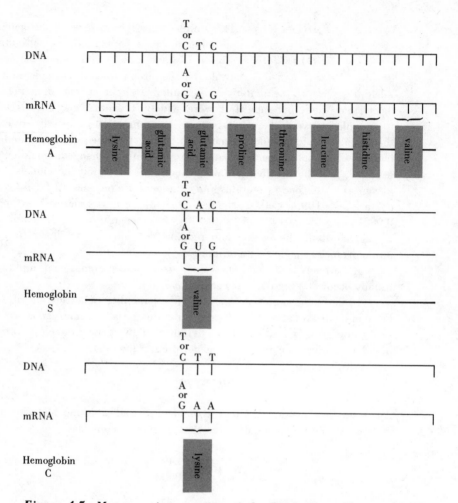

Figure 4-5 *Mutations in a portion of the DNA chain that specifies the amino acid composition of part of the human hemoglobin molecule. Hemoglobin A may mutate to produce hemoglobin S by a single change in the base in the sequence of the DNA (T → A) or hemoglobin C by a change in the base sequence (C → T). The other 103 amino acids are identical in all three hemoglobin alleles. The diagram shows the DNA and messenger RNA (mRNA) base sequences and the resulting amino acid differences in the three proteins (A, S, C).*

for μ in the above formulas, where m = the migration pressure (the proportion of immigrant alleles, q, in the total allelic pool of the population affected by immigration) and q_1 = the gene frequency of the allele in the source population.

Mutation in the simple case described above is a driving force for slight evolutionary change through the impact of continuous mutation rates. In actuality, reverse mutation $(a_2 \rightarrow a_1)$ and the other evolutionary forces counteract and modify the drive toward complete change produced by mutation, and contribute toward the maintenance of variation. Evolution may consist of nothing more than a change of gene frequencies from one generation to the next as a result of mutation pressure, but it is rarely so simple. A number of genes are undergoing regular mutational changes, recombination is mixing the genes into new gene combinations, and all the factors contributing to variation are under the impact of other evolutionary forces. The origin of variation through mutation is a central feature of evolution, but by itself it produces no long-lasting or profound changes. The other forces cannot operate without variation, but their impact far outranks mutation as the guide for evolutionary change.

THE ELEMENTAL FORCE OF MUTATION

Evolution is based on genetic variation and changes in gene frequency. The ultimate source of all genetic variation is mutation, whether produced by changes in the molecular arrangement of genes, in the linear sequence of the genes on the chromosome axis, or in the number of chromosomes. The elemental evolutionary force of mutation may directly modify gene frequency by recurrent mutation of a particular gene and produce evolutionary change. Recombination acts to enhance the effects of mutation by assembling a broad spectrum of gene combinations. It modifies and intensifies the contribution of mutation. but cannot be regarded as an evolutionary force, since it never changes gene frequencies. Recombination looms large in the evolutionary process because it provides the bulk of genetic variation that is worked on by the forces of selection and genetic drift to produce evolutionary change. Mutation provides the source of variation, and recombination an effective agent for its spread through the population. Together they develop the genetic materials for evolution.

FURTHER READING

Anfinson, C. B., *The Molecular Basis of Evolution*. New York: Wiley, 1959.

Dobzhansky, T., *Genetics of the Evolutionary Process*. New York: Columbia University Press, 1970.

Goodenough, U., and R. P. Levine, *Genetics*. New York: Holt, Rinehart and Winston, 1974.

Jukes, T. H., *Molecules and Evolution*. New York: Columbia University Press, 1966.

Levine, R., *Genetics*, 2d ed. New York: Holt, Rinehart and Winston, 1968.

Stebbins, G. L., *Chromosomal Evolution in Higher Plants*. Reading, Mass.: Addison-Wesley, 1971.

Wagner, R. P., and H. K. Mitchell, *Genetics and Metabolism*. New York: Wiley, 1964.

chapter 5

The Role of Natural Selection

The idea of natural selection as the guiding force of evolution was the principal contribution of Charles Darwin to evolutionary theory. Darwin's concept of selection was somewhat unsophisticated and negative and emphasized differential mortality of individuals rather than their reproductive contributions to the gene pool of the next and subsequent generations. Nevertheless, his recognition of this essential principle provided the key to understanding evolutionary processes. Darwin saw the process of evolution as a struggle for existence between individual organisms. Since all species produce many more offspring than survive, he concluded that the total environment eliminated those individuals least fitted for reproduction, and encouraged the survival of the fittest. The environment thus acted as a selective force, sorting out those variants best adapted to the particular environmental circumstances. Natural selection was this impact of the environment on hereditary characteristics, and its effect was descent with modification. Natural selection favored the features of an organism that brought it into a more efficient adaptive relationship with its environment, and accounted for the fact that every living creature is constructed to live in a certain environment.

The theory of natural selection has had a twisted history since Darwin's day. Complex selection was believed to be the only source of evolutionary change by many nineteenth-century biologists; it was completely repudiated by others. Among those who rejected selection as a force in evolution were early twentieth-century geneticists, confident in their new-found knowledge of heredity and certain that since Darwin had not known Mendel's principles, he could not have known very much about anything else either. It is ironic that selection today forms the only scientifically tenable solution to the problem of evolutionary change, and that its reinstatement in a modified form is principally the result of the intensive and brilliant studies of populational geneticists.

THE NATURE OF NATURAL SELECTION   Up to this point in our discussion of evolutionary forces, we have considered the mechanism of inheritance and the genetic sources of variation. We have noted that under standard environmental conditions all genes in a populational gene pool come to equilibrium and that this equilibrium is maintained. Evolution does not occur under such circumstances because evolution means change, not equilibrium.

The key to understanding evolution thus lies not in knowing the source of variation, but in finding out how the balance between various gene combinations is modified so that the composition of the population changes. One way in which small evolutionary change may occur is through mutation, but unidirectional mutation always results ultimately in a new homozygous equilibrium. The single force primarily responsible for upsetting genetic equilibrium is natural selection.

Natural selection at its simplest is the impact of any factor in the total environment of an organism that tends to produce systematic genetic change from one generation to the next. Or, stated another way, it is a relative change in the reproduction of certain genes produced by any environmental feature. Natural selection brings about evolutionary change by favoring differential reproduction of genes. Differential reproduction of genes produces change in gene frequency from one generation to the next. Natural selection does not produce genetic change, but once genetic change has occurred it acts to encourage some genes over others.

Selection is further characterized by its invariable encouragement of genes that assure the highest level of adaptive efficiency between the population and its environment. When two or more gene combinations are present, selection favors increased reproduction of the gene combination most efficient under the environmental circumstances. Evolution through selection brings about improvement in adaptive relations between organisms and their environment. Selection has been the principal force operating over millions of years to facilitate the development of new adaptations to the world's environments and is responsible for the evolution of the present diversity of life. The interaction of mutation, recombination, and selection results in new adaptive relations between organisms and their environments and forms the process of adaptation.

An experimental example of selection as a force in genetic change is provided again by the white-eyed mutant *Drosophila*. If white-eyed males are the only mates available, white-eyed and wild-eyed (normal) females will breed with them. But if the total environment is changed by including both white-eyed and wild-eyed males, one may see how selection operates (Figure 5-1). The original experimental population is composed of 50 percent white-eyes. Obviously, unless environmental conditions are important, the population will remain at equilibrium forever. In actuality,

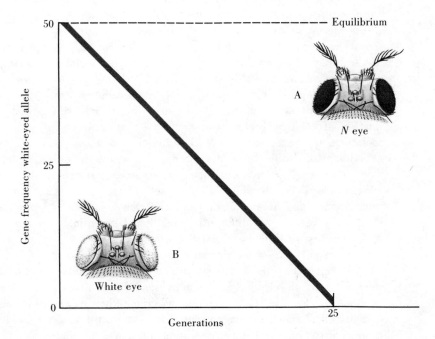

Figure 5-1   Selection for eye color in an experimental population of Drosophila.

however, both white-eyed and wild-eyed females seem to be reluctant, if not revolted, at the prospect of mating with white-eyed males (besides, the males cannot see very well) ; because males are selected differentially, the white-eye gene quickly is eliminated from the population. Mating preference thus is one factor in the total environment of this population. It acts as one selective force producing a regular systematic change in gene frequencies. In addition it produces an adaptive improvement, because wild-eyed males have a better environmental relation (that is, they are able to reproduce) than white-eyed males.

Natural selection is a creative force in evolution, since it favors and encourages efficient gene combinations. Unfortunately, many of Darwin's followers emphasized the negative aspects of selection and created the notion that it is a ruthless force, eliminating some individuals while favoring others. The idea of Darwinian selection was based principally on the concept of differential mortalities that may result in differential reproduction, but this type of selection is only one part of the total force of selecion. No white-eyed males were killed in the example given above, yet selection effectively eliminated the nonadaptive white-eye allele. Or perhaps it is better to say that selection encouraged the wild-type allele to take over the population. In any event, selection creates new adaptive relations between population and environment, by favoring some gene combinations, rejecting others, and constantly molding and modifying the gene pool. Innumerable laboratory investigations substantiate the active, positive role of selection in the modification and evolution of populations.

The workings of natural selection are exceedingly complex because of the range of organizational levels at which it functions. Selection discriminates among available reproducible biotic entities to produce more efficiently adapted units. Natural selection operates on every stage in the life history of an organism. It produces nonrandom differential reproduction of biological units and may affect any biotic entity from the molecular to the community level. Examples of levels at which natural selection makes differential discrimination are the following: intermolecule, intergene, interchromosome, intergamete, interindividual (Darwinian selection), interfamilial (kin selection), interdemic, interracial, interspecific, and intercommunity. Darwinian selection, as now understood, may result from differential viability, differential mortality, differential fertility, or differential natality, among others.

In recent years a few evolutionists have maintained the position that selection operates only at the Darwinian (interindividual) level and that the only unit of selection is the individual. In particular, this view discounts the role of differential *group selection*, where selection affects two or more members of a lineage group as a unit. Such lineage units include: a bonded-pair of adults, a set of siblings, parents and young, an extended family, a close-knit tribe of related families, a deme, a race, or a species. Well-documented examples of situations in animals where group selection easily can be envisaged include the following.

Pair-bonding—as in the deep-sea anglerfish (*Ceratias holboelli*) of the world oceans, where a small male (or males) is intimately attached as parasites to the body and connected to the bloodstream of the relatively huge female.

Food sharing—as in the Cape hunting dog (*Lycaon*) of southern Africa, where some individuals bring food back to other individuals caring for the cubs in dens.

Predator thwarting—as in the chacma baboon (*Papio ursinus*) of southern Africa, where dominant males station themselves in exposed locations while the rest of the troop feeds; if a predator approaches, the males bark and threaten the predator and cover the rear of the troop, sometimes with the aid of other males, as it retreats to safety.

Cooperative breeding—as in the American alligator (*Alligator mississippiensis*) of the southeastern United States, where several females build a common nest of leaves and decomposing vegetation on land where their eggs are deposited, and alternate guarding the nest while the other females breed.

Social colonies—as in the honey bee (*Apis mellifera*), where the entire colony cooperates to produce food, thwart predators, control hive temperatures, and insure reproduction of offspring by the queen.

These cases represent only a few of the numerous examples in animals, where group selection operates (see Figure 10-8 for other examples based on parental care in frogs). From them, it is clear that any difference in adaptive value between two group units of the same level (that is, social colony A versus social colony B) will produce a differential change in the relative frequency of genotypes (and genes) in the two groups in the next generation. In a sense, each group unit may be treated, for this purpose, as the equivalent to a complex gene combination some of which are more effective in ensuring the reproduction of their offspring than others. In extreme cases group selection may lead to survival and reproduction of one unit, while another unit of the same kind becomes eliminated or, at the level of demes, races, or even species, becomes extinct.

Groups rather than individuals may represent a less common level for the action of selective pressures, and the action of group selection may be less efficient than Darwinian selection in changing gene frequencies in a population from one generation to the next. Nevertheless it remains a significant factor in populational change, evolution, and diversification.

Another general way to look at natural selection is from the functional interrelationships between organism and environment. Selection processes may be grouped into one of the following modes based on the adaptive interplay between environment and the gene pool of a population (Figure 5-2).

CONSERVATION

INNOVATION

A. Organism—Environment relationship relatively constant.
  1. Stabilizing selection—maintains *constancy* of adaptive norm.
     a. Normalizing selection—eliminates harmful mutants, malformations, and malfunctions
     b. Canalizing selection—favors stability of gene complexes controlling development so they are not responsive to gross environmental fluctuations or to minor gene mutations
     c. Balancing selection—maintains genetic heterogeneity
B. Organism—Environment relationship dynamic; either gene pool or environment changed
  2. Directional selection—favors a regular change in one direction toward certain adaptive feature
  3. Diversifying selection—acts to break up a formerly homogeneous group into two or more adaptive norms

In the case of stabilizing selection the constancy of the environment regulates and maintains the constancy of the genetic makeup of the population. In the case of dynamic selection innovation may arise genetically (a new favorable mutation, recombination, or immigrant) that is more adaptive in the unchanged environment, or it may arise through environmental change that in turn favors a formerly nonadaptive or neutral genetic feature. Occasionally both changes occur.

In the discussion that follows only two of the many levels of selection are considered in detail. Both involve the mode of directional selection. Nevertheless, they illustrate the relation between populational selection and selection under natural conditions, and the reader may extrapolate these principles to other levels or modes of selection.

THE INTERACTION OF VARIATION AND SELECTION Because the environment is never stable, but undergoes almost constant (if frequently minor) change, the nature of selection also fluctuates. It seems probable that many of the slight changes in gene frequencies known to occur between generations in natural populations are due to changing selection pressures (see Figure 3-1). Nevertheless, on rather long-term bases the force of selection tends to channel variation along particular lines of environmental stability. From the viewpoint of populational genetics, the influence of long-term selection

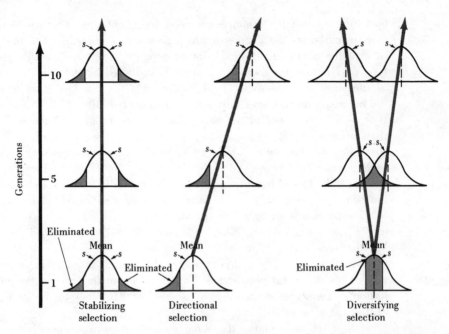

Figure 5-2 *Modes of natural selection and their effects on genetic variability and evolutionary direction. The genetic variation of each population
is represented by a normal curve. Time is measured by number of generations. In stabilizing selection, the selection pressure(s) favors the genotypes
near the mean so that the population remains genetically constant generation
after generation. In directional selection, the selection pressure drives the
population toward a different range of genotypes. In diversifying selection,
selection pressures split the original variation and drive each fragment population toward different ranges of genotypes. How do these modes of selection
relate to the patterns of sequential and divergent evolution?*

may be measured as it affects the differential reproduction of single genes.

Natural selection is the differential change in relative frequency of
genotypes due to differences in the ability of their phenotypes to obtain
representation in the next generation. If selection in a theoretical population of organisms is considered, several significant points will emerge.
Suppose that the population when first studied contains two alleles for
a particular gene (a_1 and a_2). One allele (a_1) is dominant over the other
(a_2). The gene frequencies for a_1 are ($p = 0.9$) and for a_2, ($q = 0.1$).
The population will contain three genotypes, a_1a_1, a_1a_2, a_2a_2. If the population is at equilibrium, the gene frequencies will remain constant; but
suppose that investigation reveals differential reproduction due to selection favoring the dominant allele. The effect of selection is represented by
the selection coefficient s, which indicates the force of natural selection

operating against the homozygous recessive phenotype. The selection coefficient measures the degree to which a genotype contributes to the gene
pool of the next generation. If, for example, $s = 0.2$ against this phenotype (a_2a_2), then for each 100 gametes contributed by individuals of each
of the other two genotypes (a_1a_1, a_1a_2) only 80 will be contributed by
the homozygous recessive individuals. The contribution of any one genotype to the gene pool of the next generation in relation to the contributions of other genotypes in the same population is measured by the *adaptive value* or *Darwinian fitness* (W). In this case the adaptive value for
each genotype is homozygous dominant (a_1a_1), $W = 1$; heterozygous
(a_1a_2), $W = 1$; and homozygous recessive (a_2a_2), $W = 1 - s$ or 0.8. Since
the adaptive value (W) is a relative measure, it is merely a convention
to give the value of unity to the fitness genotype; the adaptive values
might just as well be a_1a_1, $W = 1 + s$; a_1a_2, $W = 1 + s$; and a_2a_2, $W = 1$.
For this reason it is appropriate to view selection in this case as favoring
reproduction of a_1 alleles, by causing a differential increase in reproduction of the dominant phenotypes. This leads to an increase in the contribution of genotypes containing the dominant allele (a_1) to the gene pool
of the next generation.

Now let us examine the same theoretical example population, but
work with a selection coefficient $(s) = 0.1$ against the homozygous re-
cessives (a_2a_2). Again, since the two genotypes $(a_1a_1$ and $a_1a_2)$ are not
affected by selection in this case, they produce equal numbers of offspring.
As in the previous case, the genotype a_2a_2 produces fewer numbers of offspring in relation to the numbers contributed to the next generation by
the other two genotypes. The adaptive value for each genotype is a_1a_1,
$W = 1$; a_1a_2, $W = 1$; and a_2a_2, $W = 1 - s$ or 0.9. From these data the
impact of selection from one generation on to the next may be calculated
(Table 5-1).

If the same selection pressure continued indefinitely, ultimately the
recessive allele would be eliminated and equilibrium established.

The different effects of different magnitudes of selection pressures
may be illustrated in summary fashion (Table 5-2). The initial frequencies
are standardized: $p = 0.5$, $q = 0.5$. To make comparisons easy, the selection coefficients indicate the force of natural selection operating against
the homozygous recessive phenotype and genotype. Also, the fitness values
of the dominants $(a_1a_1$ and $a_1a_2)$, as in the previous examples, are unity
$(W = 1$ for each genotype). This summary provides insight into the impact that different selection pressures have on the gene frequencies of the
next generation. Different selection values produce different results in
gene reproduction, but selection, no matter how little, will ultimately
establish the favored gene.

Selection in nature rarely operates without conflict with mutation.

Table 5-1   Changes in Gene Frequency Produced by Selection Favoring the Dominant Allele

| | GENOTYPES | | | | Gene |
	a_1a_1	a_1a_2	a_2a_2	Totals	Frequencies
W	1	1	$1-s$	$\overline{W}$	p $\quad$ q
Initial frequencies	$p^2 = 0.81$	$2pq = 0.18$	$q^2 = 0.01$	1	.9 $\quad$ 0.1
Frequency after selection	p^2 (0.81)	$2pq$ (0.18)	$(1-s)q^2$ (0.009)	$1 - sq^2$ (0.999)	0.9009 0.0991

Frequency of a_1 (p) in next generation is

$$\frac{p}{1 - sq^2} = 0.901$$

Frequency of a_2 (q) $= 0.099$.

Table 5-2   Changes in Gene Frequency Produced by Different Amounts of Selection Favoring the Dominant Allele

W	0	0.4	0.9	0.99	1.5
s	1.0	0.6	0.1	0.01	-0.5^a
Initial gene frequency (p)	0.5	0.5	0.5	0.5	0.5
Frequencies after selection (p)	0.67	0.58	0.5128	0.5012	0.444
Increment of gene frequency of p	$+0.17$	$+0.08$	$+0.0128$	$+0.0012$	-0.056

^{a} The minus sign indicates that selection favors a_2 *over* a_1.

If mutation is occurring from $a_1 \rightarrow a_2$, the following formula gives the approximate change in gene frequency:

$$\Delta q = \mu p - sq^2p$$

The value μp measures the rate of mutation adding a^2 to the population, and the value sq^2, the loss of a^2 through selection. Using this formula, solve the following problem, given these values: originally $p = 0.6$, $q = 0.4$, $\mu = 0.00001$; $s = 0.001$. Calculate the gene frequencies for the next generation.

NATURAL Darwin clearly saw the importance of selec-
SELECTION IN ACTION tion as a prime evolutionary force in the
natural world. He cited innumerable examples
of adaptation through selection, and thousands of new cases are recognized each year. But one problem in discussing selection has usually been
the correlation of experimental studies under controlled laboratory conditions with what actually happens in nature. During the preceding 40
years a series of studies combining experiment and observation in the
field have been carried out on the problem of industrial melanism in
moths; they provide an exciting insight into the operation of selection
under natural conditions. These brilliant investigations were undertaken
originally by R. A. Fisher and E. B. Ford and more recently by H. B. D.
Kettlewell, all of Great Britain.

The original study was based on the peppered moth (*Biston betularia*) (Figure 5-3). These moths were well known to amateur collectors
of the nineteenth century. Up until 1845 all known specimens were light
in color, but in that year a single black moth of this species was taken
at the growing industrial center of Manchester. Presumably at that time
its highest frequency in the population was not more than 1 percent. As
time passed, more and more black peppered moths were taken, until by
1895 the black form comprised 99 percent of the Manchester population.

Figure 5-3 (Left) *Dark and light forms of the peppered moth on the trunk
of an oak at Birmingham, England.* (Right) *Dark and light forms of the peppered moth on an oak in an unpolluted area in Dorset. (Courtesy of H. B. D.
Kettlewell)*

By the early 1960s, only a few light-colored populations persisted and most areas supported populations homozygous for the black phenotype. This change in gene and genotype frequencies corresponds beautifully with the spread of industry in England during the previous 120 years. The change from light to dark color in the sooty, dirty, coal-dust-covered areas of England seems an excellent example of natural selection favoring an inconspicuous color pattern matching the darkened vegetation. Since the initial studies on the peppered moth, about 70 other species in Great Britain have been found to exhibit similar adaptive trends. The phenomenon, now called *industrial melanism,* is also known for European industrial areas, and about 100 species of moths in the Pittsburgh region of the United States exhibit a similar evolutionary change, with black forms, originally rare, taking over in conjunction with the march of industry.

Almost all evolutionists accepted the changes in the peppered moth and in other forms as a marvelous example of protective coloration being taken on rapidly as an adaptive advance, but confirmation was still lacking. Observation of the changes in gene frequency did not provide unquestioned proof of the role of natural selection in establishing industrial melanism in moths. Of course, the fact that the vast majority of forms exhibiting industrial melanism are species that rest on the surface of vegetation during the day (the exact places where soot is thickest) and which are preyed on by birds further confirmed the theory of selection. Final proof was provided by Professor Kettlewell's ingenious experiments on the significance of color in protecting the moths from their only predators, birds. Kettlewell released known numbers of marked individuals of the peppered moth into two areas: (1) a bird reserve in Birmingham, an industrial area where the local population consisted of 90 percent black moths; and (2) an unpolluted Dorset countryside, where no black moths occurred. He released 477 black and 137 light individuals at the Birmingham site. From a distance he was able to watch birds feeding on the released moths, and he recovered the surviving moths by attracting them to a light at night. The recaptures consisted of 40 percent of the black moths and 19 percent of the light moths. At the unpolluted locality 473 black moths and 496 light moths were released. The results of the recaptures were just the reverse of those at Birmingham. Only 6 percent of the black moths were retaken, but 12.5 percent of the whites. Obviously natural selection in terms of bird predators plays an enormously significant part in industrial melanism.

A further significant experiment verifies the role of natural selection in this case. At both Birmingham and Dorset, birds were filmed in the act of taking the moths from locations where equal numbers of both black and light individuals were on tree trunks. At the former site 15 black and 43 light moths were eaten; at the latter, 164 black and 26 light moths. The rapidity of change from light to dark color variants in areas of industry within natural populations of moths attests to the efficacy of natural selection as a creative evolutionary force.

Interestingly enough, during the preceding 10 years, the largely successful effort of the British government to control air pollution has led to an increasing frequency of light-colored peppered moths. Some populations previously primarily composed of black phenotypes have now reversed the trend, and light-colored moths predominate. This remarkable shift is further evidence of the effectiveness of natural selection in bringing about rapid adaptive changes in populations as the environment changes.

FURTHER READING

Bajema, C. J. (ed.), *Natural Selection in Human Populations*. New York: Wiley, 1971.

Darwin, C. R., *On the Origin of Species by Means of Natural Selection*, 1859. Numerous editions.

Fisher, R. A., *The Genetical Theory of Natural Selection*. New York: Dover, 1929.

Ford, E. B., *Ecological Genetics*. New York: Wiley, 1964.

Lerner, I. M., *The Genetic Basis of Selection*. New York: Wiley, 1958.

Muller, H. J., "The Darwinian and modern conceptions of natural selection," *Proceedings of the American Philosophical Society*, vol. 93 (1949), pp. 459–470.

Spiess, E. B. (ed.), *Papers on Animal Population Genetics*. Boston: Little, Brown, 1962.

Stebbins, G. L., "Reality and efficacy of selection in plants." *Proceedings of the American Philosophical Society*, vol. 93 (1949), pp. 501–513.

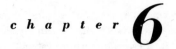
Genetic Drift

The force of natural selection acts on variation to encourage favorable gene combinations and to eliminate unfavorable ones. The
result is always the same: the development of
more efficient adaptive relationships between
a population and its environment. Selection
plays such a significant role in evolution, and
its impact on populations of organisms is so
overwhelming, that many evolutionists were
led to believe that *all* microevolutionary
changes are attributable to the interaction of
variation and selection. Concomitant with this
belief was the concept that *all* evolutionary
changes are adaptive in nature. Selection interacting with variation does account for a
great deal of evolutionary change, but recent
studies indicate that selection is not the only
force acting on variation to produce populational change.

Evidence from naturally occurring populations strongly suggests that variational differences between populations within certain species cannot have been stabilized by selection. In these population systems, various genetically regulated characters show random variation from population to population, without apparent correlation with changes in environmental (selective) factors. Numerous cases of apparently random variation between populations have been recognized by field biologists, and these cases pose a puzzling difficulty when selection is considered the only driving force in evolution. In some populations, these genetically controlled characteristics probably have a neutral selective value, yet they persist. Other characters are apparently nonadaptive but are fixed in certain populations. Typical examples are discussed in Chapter 8.

The occurrence of nonadaptive or neutral gene combinations produced in spite of selective pressure has also been demonstrated experimentally in laboratory populations. Studies in populational genetics, principally by Sewall Wright, have led to the recognition of a third elementary evolutionary force responsible for the fixing in populations of nonadaptive or neutral characteristics. This force — *genetic drift,* or the Sewall Wright effect — apparently plays an important role in populational evolution.

FIXATION OF NONADAPTIVE OR NEUTRAL CHARACTERS

When acting as a force in microevolution, genetic drift interacts with variation and selection to produce changes in the proportions of gene combinations from one generation to the next. Although the processes responsible for drift operate in both large and small populations, it is only in the latter that these processes produce a significant evolutionary effect. Genetic drift does not occur in large or moderate-sized populations, but only influences evolutionary change in small populations.

The basis for the effect of genetic drift is a matter of probabilities and operates in all populations regardless of size. To appreciate the underlying basis of this phenomenon, imagine a population containing proportional numbers of individuals of the genotypes a_1a_1, $2(a_1a_2)$ and a_2a_2, so that the gene frequencies are $p = 0.5$ and $q = 0.5$. On the basis of the Hardy-Weinberg equilibrium formula, the proportions of a_1 and a_2 will remain the same, if all other factors remain constant, in each succeeding generation. The Hardy-Weinberg formula summarizes all the individual crosses within the population. But what if a single cross between two individuals is analyzed? Will the proportions of their particular series of offspring be exactly the same as for the population as a whole? Let us take a cross between individuals of genotypes a_1a_1 and a_1a_2. If an extremely large number of offspring are born, they will often approach equal ratios of a_1a_1 and a_1a_2. However, because of the laws of probability, the ratios of genotypes will be various if the numbers of offspring are 0, 1, 2, 3, 4, 5 . . . n. If no offspring are produced or survive to adulthood, 3 a_1 and 1 a_2 genes are eliminated. If one offspring survives, the probability that the individual contains allele a_2 is 0.5; if two survive, the probability that an a_2 allele is present in both of them is 0.25; and so forth. However, all the offspring, whether 1 or 10, may contain none of the a_2 allele or they may all contain it. If one individual is produced, it may be a_1a_2 (0.5) or a_1a_1 (0.5). If 10 are produced, a number of combinations, each with a different probability, occurs: 10 a_1a_1; 9 a_1a_1 and 1 a_1a_2; 8 a_1a_1 and 2 a_1a_2; 7 a_1a_1 and 3 a_1a_2; 6 a_1a_1 and 4 a_1a_2; 5 a_1a_1 and 5 a_1a_2; 4 a_1a_1 and 6 a_1a_2; 3 a_1a_1 and 7a_1a_2; 2 a_1a_1 and 8 a_1a_2; and 1 a_1a_1 and 9 a_1a_2; 10 a_1a_2. If any situation other than 5 a_1a_1 and 5 a_1a_2 results, one allele and one genotype will be unbalanced in the genetic pool. For example, if the combination of 8 a_1a_1 and 2 a_1a_2 offspring is produced, this could change the gene frequencies and genotypic proportions in the next generation. However, in a large population, other individual crosses will produce broods containing 2 a_1a_1 and 8 a_1a_2 gene or genotypic frequencies in the next generation. In small populations the same series of processes occurs, except that the small number of individuals increases the probability that unbalanced proportions of genes will be passed on to the next generation to change the gene and genotypic frequencies and translate this random effect into genetic drift.

To illustrate this point imagine a small, homogenous population of just two individuals. A single mutation $a_1 \rightarrow a_2$ produces a single heterozygous individual. The heterozygous individual crosses only with the homozygous individual; hence, only offspring of a_1a_1 or a_1a_2 genotypes are produced. The results on populational effects in this example are directly comparable to the individual cross in the large population discussed above. The same possibilities occur, but now they may produce changes in the population. If only one offspring is produced, it may be a_1a_1 or a_1a_2, and the mutant may be lost. If 10 are produced, any one of the 10 combinations of genotypic ratios described above may form the next generation (10:0, 9:1, 8:2, 7:3, 6:4, 5:5, 4:6, 3:7, 2:8, 1:9, 0:10). If any combination except 5 a_1a_1 and 5 a_1a_2 occurs, the gene frequencies are different from those in the original population. As long as the population size remains small, random processes continue in succeeding generations so that erratic and random changes in the frequencies of the genes occur. Because the changes from one generation to the next are at random, sooner or later the frequency of a particular allele (a_2) approaches 1.0 or 0.0. If in the next generation the random effects eliminate the allele or its counterpart to bring its frequency to 0.0 or 1.0, respectively, then the population becomes homogeneous and genetically fixed with one allele

(a_2) or the other (a_1), and the eliminated allele cannot reappear without mutation. The general process of random genetic fluctuation in the gene frequencies of small populations, *which leads to loss or fixation of one gene*, is *genetic drift*. The results are random, and drift will occur regardless of selection pressure. Drift is nondirectional, and loss or fixation of a particular allele is highly probable.

In the context of population genetics the sizes of populations are expressed in terms of *effective population* size, N. N does not equal the total number of individuals in the population but only the individuals now living that will be the actual progenitors of the next generation. Although determination of N for a natural population is exceedingly complex and involves evaluation of the biology and ecology of the population, for conceptual purposes it may be thought of as the number of individuals capable of breeding or reproduction at any one time. In a situation where the other evolutionary forces are absent, Sewall Wright has determined that loss or fixation of any given allele will occur in approximately $4N$ generations. Under these circumstances, *small* populations are defined as having an $N = 10 - 100$; *intermediate* populations as having an N on the order of 10,000; and *large* populations as having $N = 100,000$ or more. Where other evolutionary forces (selection and mutation, and also immigrations) affect the population, their influences are taken into account in determining the value for N. However, for purposes of this discussion the values and ranges given above will provide sufficient approximations of effective population sizes.

As has already been indicated, accidents of sampling (the genetic luck of the draw or random sampling error) occur in all populations. In populations of large or intermediate size, accidents of sampling are effectively balanced out on a probability basis so that genetic drift toward loss or fixation does not occur.

Up to this point in this book, all discussions and examples have referred to events in intermediate to large populations. In such populations the Hardy-Weinberg equilibrium will maintain, unless the gene frequencies are affected by a change in mutation (or immigration) and/or selection pressures. In small populations, on the other hand, the process of genetic drift may affect gene frequencies in a random fashion. As an example, consider an original population with an $N = 10,000$ in which a single mutation occurred ($a_1 \rightarrow a_2$). As a result there would be 19,999 a_1 alleles and 1 a_2 allele. Without strong selection pressure favoring a_2, the mutant allele has very little chance of survival, although it may persist in the descendant gene pool at a low frequency, if it is not too nonadaptive. In a small population ($N = 10$) a similar mutation would produce a gene pool of 19 a_1 alleles and 1 a_2 allele. The difference in size allows for random processes to cause the mutant to drift to loss or fixation (the population becomes homozygous for it) within approximately 40 generations,

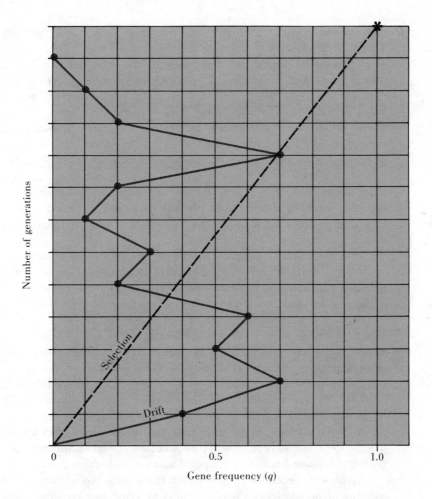

Figure 6-1 *Graphic contrast of the effects of genetic drift and selection on the frequency (q) of a single allele in two populations differing only in size (N). The solid line represents the changes in frequency in a large population where selection pressure favors the mutant allele with a frequency represented by (q) and drives the population to a homozygous state in time. The dotted line represents the changes in frequency in a small population, with the same selection pressure favoring the mutant allele as in the large population, but showing how genetic drift may eliminate an adaptive gene from a small population.*

even against strong reverse mutation or selection pressure. This point is emphasized by the illustrated example (Figure 6-1), where two populations, one small and one large, are compared. In both cases, an advantageous mutant favored by strong selection is added to the population. In the large population, selection drives the population to a homozygous condition. In the small population, with the same strong selective pressure, genetic drift operates to eliminate the favorable mutant allele.

It must be emphasized that, since genetic drift is based on random processes and not on directional ones such as recurrent mutation or selection, an allele affected by drift has an equal probability of being eliminated or fixed in the population. Thus genetic drift may enhance the effects of selection or mutation by moving the allelic frequency in the same direction favored by these forces. On the other hand, it may work in opposition to them. In short, genetic drift may operate toward elimination or fixation of a particular allele in a population against the direction of mutation and selection or in concert with them. Either possibility occurs at random and the only prediction possible is that the allele will be lost to the gene pool, *or* the gene pool will become homozygous for it within approximately $4N$ generations.

An interesting example of drift in laboratory populations demonstrates its effect. A series of laboratory populations composed of eight individual fruit flies were established. The populations were each composed of four males and four females; two of each were homozygous for the nonadaptive mutant forked wing (a_2) and for normal wing (a_1). Initial gene frequencies were a_1 $(p = 0.5)$ and a_2 $(q = 0.5)$. From the several hundred progeny produced by the original stock, eight were taken at random to form the next generation of the population. The same process was followed for 16 generations, and the population size was maintained at eight. Ninety-six populations were studied in this fashion. At the end of the experiment only 26 of the populations contained both gene a_1 and a_2. In 29 populations forked wing (a_2) had become fixed, and these populations were homozygous; in 41 populations the gene for normal wing (a_1) had become fixed, and the populations were homozygous.

Although genetic drift provides a theoretical explanation of random patterns of populational variation, no convincing experimental studies of drift in natural populations have appeared. As a result some evolutionists reject the view that genetic drift is a significant evolutionary force. Others object to its use in explaining random variation because some populations having fixed nonadaptive or neutral features are of large size. It seems likely that genetic drift may be more of a force in evolution than is currently suspected, because all populations tend to undergo fluctuations in size over relatively short periods.

Drift seems to play an effective role in evolutionary change in small populations as follows.

1. Continuous drift—where the size of a population remains small and the random sampling error each generation forms a basis for continuing genetic drift.

2. Intermittent drift—where the size of the population is occasionally

reduced, and during the time of size reduction, for several generations, drift operates to fix certain genes, including nonadaptive or neutral ones; later, when the population size increases, it may retain the homozygous genetic makeup established earlier by drift, although drift does not currently operate in the population.

    3. Founder principle—where a new population (that is, on an island) is established by a few individuals (perhaps a single gravid female) that represent only a random fraction (a random sampling error) of the genetic variability of the original parental population and sample, drift may produce a rapid differentiation from the parental stock, including the loss of adaptive genes and the fixation of nonadaptive or neutral ones.

    Demonstration of genetic drift as an evolutionary force operating in natural populations is a challenging problem needing intensive study.

    Genetic drift, although it may operate to accelerate the impact of mutation and selection pressures, is of great significance evolutionarily, since it provides a genetic basis for the establishment of nonadaptive or neutral characters in a population.

    The essential feature of drift is that the smaller the population, the greater the random variations in gene frequencies from generation to generation. In a small population the random effects causing drift lead to loss or fixation of alleles; the significant result of drift is that a homozygous population is produced within a few generations. The process is fickle. It may result in a generation having a preponderance of one allele being followed by a generation with the other more abundant; but loss or fixation of one or the other is so highly probable as to be inevitable. Since most species of organisms consist of partially isolated populations that are small in numbers at least at some time in their history, the potential role of drift in microevolution is substantial.

NEUTRAL
MUTATION-RANDOM
DRIFT HYPOTHESIS

Recently a number of molecular and population geneticists (particularly James Crow, Thomas Jukes, and M. Kimura) have developed a concept of evolutionary change in which neutral mutations and genetic drift are thought to predominate over natural selection. This idea is sometimes called the neutralist or non-Darwinian theory of evolution, although at present its validity is equivocal. The increasing interest in this idea stems primarily from an understanding of molecular genetics. As you will recall from the discussion of DNA transcription and translation (Chapter 1) and gene mutation (Chapter 4), the change in one nitrogenous base in a triplet may produce a change in the amino acid added at a particular point in the growing polypeptide chain. Often these slight differences in amino acid sequence seem to occur in an unimportant portion of the protein molecule and have no functional significance. For example, one worker discovered 32 protein variants of the xanthine dehydrogenese gene in *Drosophila,* all of which are very similar in structure and function but differ in minor details of the amino acid sequences.

The essential ingredients of the Neutral Mutation-Random Drift hypothesis center around the following viewpoints.

1. The level of variation at enzyme loci of genes in natural populations is very high; thus there is high polymorphism in the number of slightly different proteins whose production is regulated by these loci.

2. The variant proteins produced by these differences are functionally identical, so the several different proteins differ only in a minor way and are selectively neutral.

3. The rate of amino acid substitution for any one protein over time is remarkedly uniform, and such constant rates indicate a constant process of mutation unaffected by natural selection.

4. The process of genetic drift explains how these slightly variant alleles are fixed in populations.

5. Since these kinds of neutral mutations are the rule rather than the exception, most genetic variation and its patterns in natural populations are the result of neutral mutation combined with random drift rather than natural selection.

These views have been vigorously attacked by many other biologists. Particularly at issue is the conclusion that most of the amino acid substitutions in the polymorphic mutant proteins are neutral and do not affect function. In a number of cases previously thought to support the neutralist view, specific functional differences between the proteins have been demonstrated. The idea that amino acid substitutions in a given protein occur at a constant rate has also been challenged. Although the most recent experimental work seems to have decreased the role of the neutralist position as a major determinant of long-term evolutionary change, the importance of genetic drift seems firmly established as a random force in populational change.

FURTHER READING

Ayala, F. J., "Biological evolution: natural selection or random walk?" *American Scientist*, vol. 62 (1974), pp. 692–701.

Dayhoff, M. A., "Computer analysis of protein evolution," *Scientific American*, vol. 221 (1969), pp. 87–95.

Dobzhansky, T., and O. Pavlosky, "An experimental study of interaction between genetic drift and natural selection," *Evolution*, vol. 11 (1957), pp. 311–319.

Kerr, W. E., and S. Wright, "Experimental studies of the distribution of gene frequencies in very small populations of *Drosophila melanogaster*. I. Forked," *Evolution* (1954), pp. 172–177.

Kolata, G. B., "Population genetics: reevaluation of genetic variation," *Science* (1974), pp. 452–454.

Lewontin, R. C., *The Genetic Basis of Evolutionary Change*. New York: Columbia University Press, 1970.

Li, C. C., *Population Genetics*. Chicago: University of Chicago Press, 1955.

Wright, S., "Evolution in Mendelian populations," *Genetics* (1931), pp. 97–159.

chapter 7

The
Result
of Evolution:
Adaptation

The end product of evolutionary change is the establishment of organisms that function more efficiently in a certain environmental situation than did their predecessors. At the populational level the acquisition of better adaptive relations to a changing environment is accomplished primarily through the interaction of mutation and selection, but other factors may enhance the effect of these forces, and genetic drift may be antiadaptive. In evolution above the population level, another complex of forces contributes to the development of more efficient relationships between organisms and their environments.

Any characteristic that is advantageous to a particular organism or population is called an *adaptation*. The lungs of land vertebrates make possible gaseous exchange between these organisms and the air and are an adaptation for terrestrial life. The reality of adaptation is confirmed by the diversity of life and the thousands of environmental situations inhabited by life. The millions of different adaptations found among living organisms—from the photosynthetic pigments of plants, the webs of spiders, and the roots of trees, to the complex locomotor devices of fish, bats, and horses—provide us with the fact of evolutionary change. The great contribution of Jean Baptiste Lamarck to evolutionary thought was his careful demonstration of the universality of adaptation in living organisms and his recognition of adaptation as the result of evolutionary change.

The acquisition of adaptive features through the interaction of evolutionary forces is the process of adaptation. Lamarck's attempt to explain the process of adaptation was incorrect and for this reason his contributions to evolutionary thought are usually overlooked. His demonstration of the fact of evolution and the reality of adaptation was finally vindicated by Darwin, who developed a more cogent explanation of the process of adaptation.

ADAPTATION AND ENVIRONMENT Most books about evolution devote much space to presenting evidence for evolution. Invariably the argument consists of the enumeration of adaptations, usually different adaptive functions of the same structure or fossil histories of their structural development. The universal existence of adaptation need not be belabored here. The fact that organisms are alive and reproduce offspring of the same kind is in itself proof of adaptation. Any organism, including the reader of this book, is marvelously adapted for existence on earth, and each species exhibits an impressive series of general and special adaptations for life in a particular environment. General adaptations are usually most important in the long-term evolution of a major group of organisms; special adaptations are developed for restricted and specialized adaptive relations to a small segment of the available environment.

The sand lizard (*Uma scoparia*) of the Mojave Desert of California (Figure 8-1) exhibits both general and special adaptations for life on desert sand dunes. The special adaptations, advantageous for dune life, include fringed toes, shovel nose, valvular nostrils, modified eyelids, color pattern, and the behavior of swimming under the surface of the sand. Less obvious, but perhaps even more significant to life on windblown dunes, are the general adaptations of the camera eyes, lung, circulatory system, digestive tract, nervous system, and many other features found in all lizards and most terrestrial vertebrates.

Adaptation has produced living forms in tune with the requirements of their total environment. The complexity of the total environment explains the complexity of adaptation and the necessary compromises between adaptations to conflicting environmental forces. All living organisms live in an environment comprised of the physical and biotic influences of their particular ecosystem and are adapted in an extremely complex way to the matrix of interacting factors:

A. Physical environment: physical and chemical factors
B. Biotic environment
   1. Extraspecific environment: biotic community
   2. Interspecific environment: population
   3. Internal environment: individual

THE PROCESS OF ADAPTATION   In preceding chapters the several forces responsible for evolutionary change—mutation, selection, and drift—have each been considered as more or less independent elements. In actuality, all three forces operate contemporaneously in most populations. The basic process of microevolution consists of changes in gene frequencies in a population, from one generation to the next, as guided by the elemental forces. The forces upset the genetic equilibrium to produce microevolutionary change.

The elementary process of evolutionary adaptation involves microevolutionary change brought about by the interaction of variation and selection (Figure 7-1). Selection operates on the available variation to produce a gene pool that interacts more efficiently with the particular environment than did the preceding population. Any change in variation through mutation or recombination provides new grist for the evolutionary mill as the grindstone of natural selection sorts out some gene combinations and favors others. A complicating factor in this simple pattern is genetic drift, which may also affect variation but does not usually favor adaptation.

As we have discussed the elemental forces, emphasis has been placed on single genes for the sake of explanation. In all known cases of microevolution, however, many genes and gene combinations are under the influence of hundreds of environmental factors. Changes in gene frequency are rarely as simple as our examples imply. Even at its simplest, evolution is an extremely complicated process. Selection molds the spectrum of variation into new patterns of adaptation in a continuously changing environment, resulting in change in the environment-organism relations.

Microevolution through the interplay of the three elemental forces of evolution probably produces changes in all populations from one generation to the next. Microevolution is also responsible for the differences that arise between related populations. Additional evolutionary forces usually

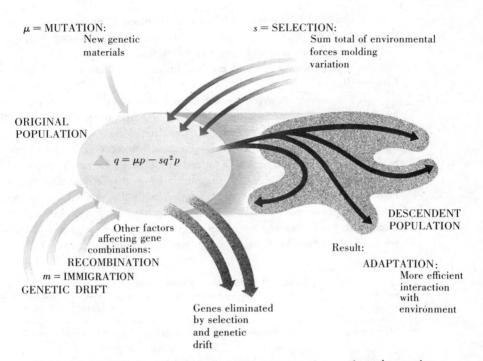

μ = MUTATION:
     New genetic
     materials

s = SELECTION:
     Sum total of environmental
     forces molding
     variation

ORIGINAL
POPULATION

$q = \mu p - sq^2 p$

Other factors
affecting gene
combinations:
RECOMBINATION
m = IMMIGRATION
GENETIC DRIFT

Genes eliminated
by selection
and genetic
drift

DESCENDENT
POPULATION

Result:
ADAPTATION:
     More efficient
     interaction
     with
     environment

Figure 7-1 *The elementary evolutionary process: interaction of mutation, recombination, selection, and drift to produce adaptation. The area of the original population symbolizes variation in gene combinations produced by mutaton, recombination, and drift; the area and shape for the descendant population symbolize the new range of variation in gene combination produced by gene elimination through selection and drift.*

cooperate with mutation, recombination, selection, and drift to produce new populations from preexisting ones, and the process of adaptation is usually involved (see Figure 7-1). The fragmentation and development of new populations is called *speciation* and will be considered in detail in Chapter 9.

At another level, fragmentation involves the origin and evolution within a short time of a great many adaptive types. This pattern of fragmentation may include a number of species populations but usually occurs above the species level. It is called *macroevolution*, or *adaptive radiation*. Macroevolution is characterized (1) by subdivision of the group into many new subgroups, (2) by an invasion of numerous new environmental situations, and (3) by diversification of structure and biology. While microevolution and speciation tend to produce special adaptation, macroevolution usually develops from a general adaptation with a number of special adaptations following from the general one.

Finally, on rare occasions, new combinations of characteristics cause the appearance of new biological organizations of general adaptation. The evolution of these new and rare organizational adaptations forms *megaevolution*. Macroevolution and megaevolution are discussed more fully in Chapter 10.

All the levels of evolution differ to a considerable degree from one another in fundamental features, but all are based upon the microevolutionary process and all contribute to adaptation. Microevolution alone produces sequential adaptive change; speciation, macroevolution, and megaevolution produce divergent adaptation. At all levels, the result of evolution is the same: the development of organisms adapted to a changing environment and having a more efficient relation with the present environment than their predecessors. With the sole exception of microevolutionary genetic drift, the forces and processes of evolution lead to higher and higher levels of adaptation between organism and environment (Table 7-1).

Table 7-1 *Levels of Evolutionary Change*

Process	Basic Pattern	Result
Microevolution	Sequential	Adaptive, neutral, and nonadaptive changes in gene combinations
Speciation	Divergent	Adaptive changes in isolated related populations
Macroevolution	Divergent	Adaptive diversification and radiation
Megaevolution	Divergent	Origin of major new adaptive biological organizations

FURTHER READING

Carlquist, S., *Island Biology*. New York: Columbia University Press, 1974.

Cott, H. B., *Adaptive Coloration in Animals*. London: Methuen, 1940.

Dobzhansky, T., *Evolution, Genetics and Man*. New York: Wiley, 1955.

Ehrlich, P. R., and P. H. Raven, "Butterflies and plants," *Scientific American*, vol. 216 (1967), pp. 104–113.

Grant, V., *The Origin of Adaptations*. New York: Columbia University Press, 1963.

Marshall, N. B., *Aspects of Deep Sea Biology*. London: Hutchinson's, 1954.

Simpson, G. G., *The Major Features of Evolution*. New York: Columbia University Press, 1953.

Wilson, E. O., *The Insect Societies*. Cambridge: Belknap Press, 1971.

EVOLUTIONARY DIVERGENCE

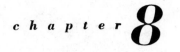

Races, Species, and Isolating Mechanisms

Anyone who has some familiarity with the living world realizes that organisms are not uniformly distributed over the earth's surface. Even if we study a single species—for example, the sand lizard (*Uma scoparia*) of the Mojave Desert in California, which has a rather restricted habitat of windblown sand dunes—it soon becomes apparent that the individuals are not evenly distributed over the species range. Further investigation will demonstrate that even on a single sand dune the individual lizards are not evenly dispersed but tend to occur in irregular clusters (Figure 8-1). Detailed study reveals that each of these small groups of individuals acts as a distinct unit. The sand lizards within the unit interact with one another, select their mates from adjacent individuals in the group, and rarely

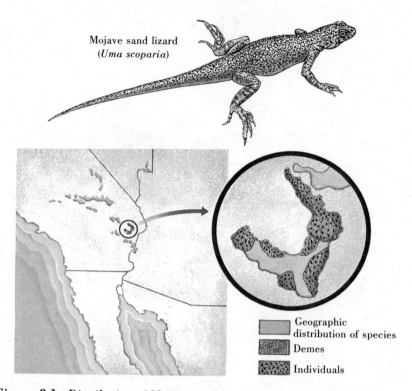

Mojave sand lizard
(*Uma scoparia*)

Geographic
distribution of species

Demes

Individuals

Figure 8-1 *Distribution of Mojave sand lizard,* Uma scorparia. *The map indicates the basic geographic distribution of the species and the insert the distribution of individuals (dots) and demes (shaded areas) on a single sand dune. Note the modifications for sand life in this species: countersunken jaw, shovel snout, valvular nostrils, specialized eyelids, protective scales over ear opening, fringed toes for locomotion on shifting sand, and general concealing coloration.*

wander across the intervening gap to associate with lizards from other such groups. Additional investigation indicates that members of each group resemble one another very closely in genetic features and differ slightly from the other groups on the same dune. Each of these clusters of individuals forms a population partially isolated both spatially and genetically from other similar populations.

These groups of genetically similar individuals bearing an intimate temporal and spatial relation to one another form the smallest population units, called *demes*. Each deme is isolated to some extent from adjacent demes, but genetic exchange is always possible between them. Even distantly located demes may contribute genetic material to one another over a period of time by the gradual passage of gene combinations from one deme to another. Demes are open genetic systems that are affected by gene flow from adjacent populations; that is, they are only partially isolated populations.

The largest populational unit is called a *species*. This kind of population is usually made up of several to many demes, and although each deme is partially isolated from other demes within the species population, there is always some potential for genetic exchanges. Species populations, however, unlike the individual demes of the same species, are closed genetic systems, protected against gene flow from other species through complexes of genetically regulated isolating mechanisms.

In the basic pattern of sequential evolution, there is almost constant change in genetic constitution within a population from generation to generation. The interaction of variation, selection, and drift, functioning within the limits of the population structure, is responsible for these continuing and gradual changes. It is almost as though the three evolutionary forces were struggling for control of the population. In one generation variation may gain the upper hand as the effects of selection and drift are reduced by population size or a changing environment. Selection may be most efficacious in the next generation, eliminating to a considerable degree the gains made by variation. Drift, of course, usually operates to the detriment of both selection and variation. At any one time the genetic composition of the population is a compromise between the effects of the three forces. Over a period of time a population may undergo enough change to become distinct from the original ancestral deme. However, the kind of evolution producing changes between generations of the same population does not in itself explain the origin of new populations.

The major feature of organic evolution is the production of new adaptive types through a process of populational fragmentation and genetic divergence. In the remaining chapters we will analyze the manner in which the elemental forces of evolution produce new populations from the old. The forces involved are fundamentally the same as those operating in sequential evolution, reinforced by additional factors to bring about evolutionary divergence.

VARIATION IN BIOLOGICAL POPULATIONS

The diversity of living organisms would seem to demonstrate that evolutionary divergence does in fact occur. The manner in which diversity arises, however, is only understandable if we appreciate the patterns of variation found in biological populations. We know that evolution occurs at the population level, and an understanding of variation at this level provides a basis for understanding how evolutionary divergence comes about.

We have already mentioned the general nature of the largest populational unit, the species. At this point it is appropriate to present a more formal definition as a basis for discussion of intraspecific variability. We know that a species is a population usually made up of a number of genetically similar demes that replace one another spatially or ecologically. The demes within the species are only partially isolated from the other demes and may exchange genetic materials with one another when in contact, as when an individual from one deme wanders into the area of another deme during the breeding season. In addition we know that all demes of a particular species are genetically isolated from demes making up other species. A species may be characterized as follows: a natural population in which the individuals are actually or potentially capable of breeding with one another and, under natural conditions, do not normally or successfully interbreed with individuals of other populations.

We recognize two patterns of isolation between related species. In some cases, two very similar populations may be widely separated from one another by a geographic or ecologic barrier. For example, the southern elephant seal, *Mirounga leonina,* occurs in the cool waters of the Southern Hemisphere around Antarctica, the southern coasts of South America, South Africa, Australia, New Zealand, and many of the antiboreal islands. A close ally, the northern elephant seal, *Mirounga angustirostris,* is found in cool waters along the coast of western North America. The two forms are very similar to each other and can be distinguished only with difficulty. However, the breeding populations of the two forms are separated by about 3,000 miles of warm tropical seas, and hence are not capable of genetic exchange. Where forms occupy discrete geographic or ecologic ranges separated from each other by spatial barriers, they are called *allopatric populations.* The two kinds of elephant seals are allopatric species isolated by an ecogeographic barrier.

The second pattern of isolation occurs in cases where related populations share a portion of their ecologic ranges. Such populations are called *sympatric* and they are isolated from each other not by space but through the physiological expression of genetic difference. The salamanders of the genus *Taricha* exemplify this situation. *Taricha torosa,* the California newt, is found from southern California northward into northcentral California. The Pacific newt, *Taricha granulosa,* occurs from southern Alaska south along the Pacific coast to the San Francisco Bay region. The two populations are sympatric in central California, both to the north and to the south of San Francisco. Even though these newts may breed at the same time and in the same streams, genetic exchange between them is prevented by differences in breeding behavior, egg deposition, and developmental patterns. Allopatric populations may be isolated from each other by physiological differences as well as by distance, but the essential significance of genetic isolating mechanisms is expressed only in sympatric species.

Within species populations there are a number of characteristic variation patterns. These patterns result from the activity of the elemental forces of evolution, and description of the patterns forms a basis for the analysis of the origin of evolutionary divergence. For our purposes the following variational patterns will be considered.

A. Random variation
B. Nonrandom variation (ecologic variation)
   1. Races
   2. Clines

Among many species of plants and animals, variation appears to form a random pattern. In these cases each deme within the species population differs slightly from adjacent demes. When the distribution of the entire species is viewed, the variation presents a chaotic pattern having no discernible correlation with differences in the environment in the various areas within the species range. A typical example of random demic variation is provided by the ocellated klipfish, *Gibbonsia elegans*, a common fish along the coast of western North America. Each deme in this species exhibits its own combination of peculiarities apparently unaffected by the characteristics of immediately adjacent demes. One character, body depth, shows a fascinating spectrum of variability, from a population with an average body depth of 16.9 percent of the body length, to one with an average body depth of 20.4 percent of the body length. If these two demes were considered alone, the individuals could always be recognized as coming from one deme or the other on the basis of body depth. Other demes are intermediate between these two extremes, and although they are all partially isolated from one another, some genetic exchange occurs. Significantly, the body-depth character exhibits a random pattern of ecogeographic distribution within the species. In this variation pattern, the genetic differences between the several demes do not seem to be correlated with habitat differences.

In many other organisms variation is nonrandom and tends to be closely correlated with differences in the ecologic conditions in different portions of the species range. In many cases the species appears to be composed of groups of essentially similar demes that inhabit large portions of the total range. The demes within the several segments usually share a large number of basic characteristics. When the segments appear to be different in genetic makeup and are readily recognized, they are called *races* or *subspecies*. Demes found in the areas intermediate to the ranges of particular subspecies usually exhibit characteristics of both adjacent forms. These intermediate populations form a zone of *intergradation* or genetic flow between the races. The western rattlesnake, *Crotalus viridis* (Figure 8-2), is an excellent example of ecologic fragmentation.

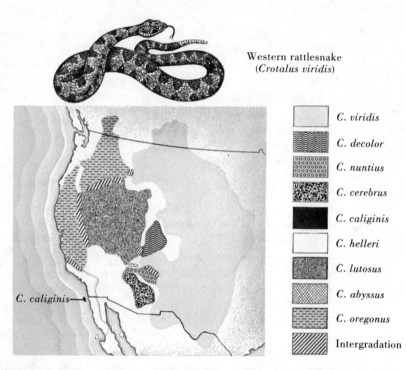

Western rattlesnake
(*Crotalus viridis*)

C. viridis

C. decolor

C. nuntius

C. cerebrus

C. caliginis

C. helleri

C. lutosus

C. abyssus

C. oregonus

Intergradation

C. caliginis——

Figure 8-2   Geographic variation in the western rattlesnake, Crotalus viridis.

Nine geographic races are recognized, and each race occupies a distinctive ecologic area within the total range of the species. The races differ in characteristics of body proportions, scutellation (arrangement of scales), and coloration. Intergrading populations are found whenever any two races of this rattlesnake come into ecologic contact.

Frequently, the ecologic forces responsible for divergence operate at a local level to produce a great many slightly different races. The western pocket gopher, *Thomomys bottae*, is a particularly plastic form capable of adaptation to a wide spectrum of conditions in soil, food supply, vegetational cover, temperature, and humidity. In consequence a large number of ecologic races within the species are readily recognizable on the basis of pelage (body coat) difference, size, and skeletal proportions (Figure 8-3). Each such population is adapted for existence in a very small ecologic segment of the total range and forms a distinctive, partially isolated ecologic race.

Another interesting aspect of the general phenomenon of racial variation is the tendency for certain rather similar adaptive types to develop independently throughout the geography of the species. In the case of the pocket gophers, populations living on the same general soil types resemble one another very closely in color of pelage, even though their ranges may be hundreds of miles apart. These populations occur under similar ecologic conditions and their similarity suggests that the basic features held

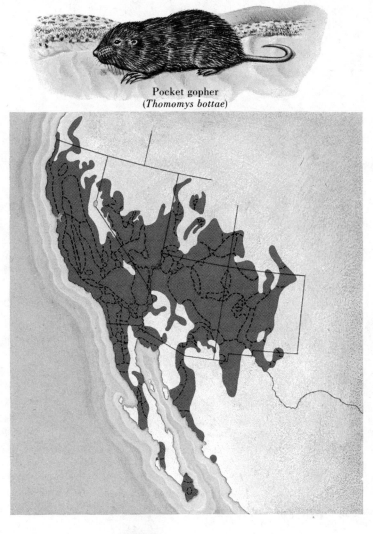

Pocket gopher
(*Thomomys bottae*)

 Species distribution

------ Races

Figure 8-3   *Ecologic fragmentation of races in the western pocket gopher,* Thomomys bottae. *Numerous local races have developed in this species in response to local ecologic conditions.*

Hawkweed
(*Hieracium umbellatum*)

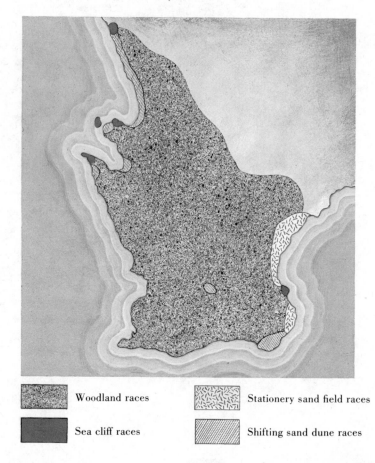

	Woodland races		Stationery sand field races
	Sea cliff races		Shifting sand dune races

Figure 8-4 Distribution of the hawkweed, Hieracium umbellatum, *in Sweden, demonstrating ecologic parallelism in sand dune populations.*

in common represent adaptations for the same kinds of environmental circumstances.

An additional example of the parallelism of ecologic adaptation is provided by the hawkweed, *Hieracium umbellatum*, of Sweden. In this plant, populations occur in four principal habitats: shifting coastal sand dunes, sandy fields, sea cliffs, and woodlands. Throughout the entire species range, demes established on coastal sand dunes resemble one another in basic adaptive features. Other adaptations are shared by all sandy field populations, others by those living on sea cliffs, and still others by woodland populations. The sand dune populations are separated from one another by populations of the woodland type (Figure 8-4), but have apparently responded to similar selective influences. In characteristics of little significance to sand dune life, populations in the various sand dune regions may differ markedly from one another.

In the above situations the pattern of demic variation exhibits a pronounced correlation with environmental conditions; sudden and obvious changes in populational characteristics can be noted between ecologic regions. The distinctive areas occupied by each ecologic race are discrete, and areas occupied by intermediate demes are very narrow. In many other species, on the other hand, the distinctive characteristics of the

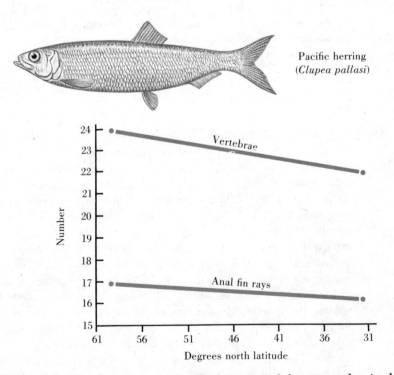

Figure 8-5 *Clinal variation in vertebral counts and fin ray number in the Pacific herring,* Clupea pallasi, *along the coast of western North America.*

demes form a continuous variational gradient correlated with environmental gradients. In such cases the changes from one deme to the next are
so slight that no obvious break is apparent in the continuum of variation.
An example is found in the populations of Pacific herring (*Clupea pallasi*)
of the eastern Pacific Ocean, which differ from one another in vertebral
number and number of fin rays. These characteristics form a continuous
variation pattern from north to south along the coast of western North
America. In Alaskan waters the populations of herring have high vertebral
and fin ray counts (Figure 8-5). Southward along the coast each succeeding population is made up of individuals with slightly fewer vertebrae
and fin rays than the population just to the north. At the extreme southern
limit of the range all individuals have fewer vertebrae and fin rays than
any individuals in the far north. While the Alaskan and Californian populations differ markedly from one another in these features, a complete
transition in vertebral and fin ray number is represented in intermediate
populations. A gradual, as distinct from a sudden, change of this kind is
called a *character cline*. The extremes in the cline are distinctive, but the
change from population to population forms so gradual a continuum that
it is impossible to group the demes into separate races. Clinal variation is
common among living organisms and is usually correlated with an ecological gradient. For example, the character clines in the Pacific herring
are obviously correlated with the water temperature gradient running from
Alaska (area of cold waters) down to southern California (area of warm
waters).

ISOLATING MECHANISMS   Races and demes within a species retain basic
genetic characteristics that allow interbreed-
ing when individuals come into contact, whereas species are isolated gene
pools. Some species cannot exchange genetic material because they are
separated spatially. Others maintain their hereditary integrity, even when
sympatric, through the operation of special isolating mechanisms that prevent interspecific interbreeding or reproduction. In most instances isolating
mechanisms serve as external barriers to reproduction between breeding
individuals of related species. In some cases individuals of two distinct
species may cross, and their hybrid progeny, with genetic characteristics
of both parents, may survive to adulthood. In these relatively rare individuals, the genetic contributions of the two parents appear to clash and
tend to produce a number of internal barriers to genetic amalgamation
that eliminate the hybrid from the breeding populations of both parental
species. The various kinds of isolation between related species are indicated in the following outline.

A. Spatial isolation: allopatric populations
B. Genetic isolation: allopatric and sympatric populations
   1. Premating isolating mechanisms — external reproductive isolation

EXTERNAL BARRIERS

      a. Ecologic — mates do not meet
      b. Ethologic — mates do not meet
      c. Morphologic — mates meet, but no gametes are transferred
   2. Postmating isolating mechanisms — internal reproductive isolation

INTERNAL BARRIERS

      a. Gametic mortality — no fertilization
      b. Zygotic mortality — fertilization occurs, zygote dies
      c. Hybrid inviability — offspring with reduced viability
      d. Hybrid sterility — offspring viable but sterile
      e. Hybrid breakdown — first generation offspring viable and fertile, later generations inviable or sterile

Concrete examples of the operation of these barriers to genetic intermixing may be readily observed in numerous species. In the area around New Orleans, Louisiana, almost all the different types of isolating mechanisms are seen to function among the various sympatric populations of frogs and toads. Each mechanism produces the same result: basic genetic characteristics are retained, and there is no contamination by gene combinations from other species.

Ecologic isolation is well illustrated by the pig frog (*Rana grylio*) and the gopher frog (*Rana areolata*). The former is extremely aquatic and occurs in deep ponds, lakes, and marshes among lily pads and emergent vegetation. The latter species occupies mammal and tortoise burrows during the day, but is active at night around the margins of swamp areas. The pig frog breeds in deep water and has no ecologic contact with the gopher frog, which breeds in isolated grassy ponds in shallow water. Differences in ecologic preference eliminate possible matings between the two species.

In many cases species with similar ecologic requirements breed at the same time and place, but other isolating mechanisms prevent genetic mixing. Near New Orleans, the closely allied gray tree-frog (*Hyla versicolor*) and pine woods tree-frog (*Hyla femoralis*) frequently breed in the same ponds. The principal barrier to interbreeding appears to be behavioral or ethologic. Female frogs and toads locate males by the calls given by the latter after they reach the breeding site. Although the gray tree-frog and the pine woods tree-frog are very similar in most characteristics, their male breeding calls are extremely different. In the gray tree-frog the call is a short trill, loud and resonant, of no more than three seconds duration. The call of the pine woods tree-frog consists of a series of short, sonorous dots and dashes. Female tree-frogs distinguish between these calls and no mixed matings occur.

Differences in size prevent hybridization between the oak toad (*Bufo quercicus*) and the Gulf Coast toad (*Bufo valliceps*). The oak toad is a minute amphibian attaining a maximum length of $1\frac{1}{4}$ inches in females. Small adult male Gulf Coast toads are $2\frac{1}{4}$ inches in length. Size alone prevents interbreeding: male oak toads are much too small to grasp female Gulf Coast toads, and male Gulf Coast toads are so large that they are more likely to eat female oak toads than to try and breed with them. Even if they were to engage in courtship, a female oak toad would drown under the weight of a male Gulf Coast toad.

Internal barriers have significance as isolating mechanisms only after other restrictions on interspecies hybridization fail. The main reason that the bronze frog (*Rana clamitans*) and its close relative the bullfrog (*Rana catesbeiana*) do not hybridize is because of a cytological block to fertilization. Even when artificial crosses, using sperm and egg preparations obtained in the laboratory, are attempted, no development occurs. Obviously the chromosomes of the two species are incompatible and will not function properly to initiate development if brought into contact with each other. In other cases, although the genetic compositions of the parental species are antagonistic, development proceeds for some time in the hybrids. The resulting clash between the diverse genetic makeups is delayed frequently, but leads to weakness or early death in the hybrid. For example, hybrids between the bullfrog and the gopher frog pass successfully through the early stages of development, but die before reaching the tadpole stage.

Other hybrids attain adult characteristics but the chromosomal incompatibilities produce gonads that fail to develop properly, and the hybrids are sterile. Hybrids between the Gulf Coast toad and Fowler's toad (*Bufo woodhousii*) are produced rather frequently under natural conditions. All individuals derived from crosses between females of the former species and males of the latter die early in development. Embryos produced by crosses between female Fowler's toads and male Gulf Coast toads develop normally into adults. The adults are all males and completely sterile. In many organisms hybrid nonviability or sterility does not appear until the hybrid produces offspring by backcrossing to one of the parental species.

In this chapter we have sketched the basic outlines of populational diversity and isolation. The essential questions concern the way in which the elementary evolutionary forces translate their effects into the origin of new populations. How does populational diversity arise? How do divergent populations become genetically isolated from one another? These questions have puzzled all students of evolution from Lamarck and Darwin to the present day. In the following chapter we will attempt to explain current ideas of the processes by which populational divergence develops within a species to form new demes and races, and how the same processes operate on the fragment populations until some of them become sufficiently isolated to form new species.

FURTHER READING

Blair, W. F. (ed.), *Vertebrate Speciation*. University of Texas Press, 1961.

Brown, W. L., and E. O. Wilson, "Character displacement," *Systematic Zoology*, vol. 5 (1956), pp. 157–161.

Ehrlich, P. R., R. W. Holm, and P. H. Raven, *Papers on Evolution*. Boston: Little, Brown, 1969.

Mayr, E. (ed.), *The Species Problem*. Washington, D.C.: American Association for the Advancement of Science, 1957.

Simpson, G. G., "The species concept," *Evolution*, vol. 5 (1951), pp. 285–298.

Stebbins, G. L., "The role of hybridization in speciation," *Proceedings of the American Philosophical Society*, vol. 103 (1959), pp. 231–251.

———, "The comparative evolution of genetic systems," in *Evolution After Darwin* (S. Tax. ed.), Chicago: University of Chicago Press, 1960.

chapter **9**

The Origins of Species

The entire course of evolution depends upon the origin of new populations having greater adaptive efficiency than their ancestors. The study of populational divergence (speciation) is therefore crucial to an understanding of evolution. At one time, and among some evolutionists even at present, the process of speciation was thought to be synonymous with evolution. As most biologists now realize, evolution is not a simple process involving only the origin of divergence. The factors responsible for race and species formation are substantially different from those operating to produce sequential microevolution, and evolution above the species level also entails a different complex of processes. Evidence has accumulated to the point where it is no longer proper to speak of a single origin of species, since we know that species may originate in several very different ways and under a variety of influences. The problem remains an intriguing one for all biologists, because we are far from fully comprehending the multiplicity of interacting factors that produce speciation in different groups of organisms.

DEME The several patterns of random and nonran-
AND dom variation exhibited within species popu-
RACE FORMATION lations immediately suggest that divergence
has different origins under differing circumstances. The simplest conceivable situations leading to the formation of new populations from old are those in which a single deme becomes established in a new habitat. Imagine that a small breeding population of snails is introduced onto a moderate-sized island (one about the size of Long Island in New York). The population comes to inhabit a relatively small area near its point of introduction and after a few generations reaches genetic equilibrium. In what ways may new demes be formed from the initial population?

Obviously, new demes may arise in this situation only through migration and fragmentation. The original population will probably tend to expand and to spread out from the center of introduction into surrounding areas. In the beginning, genic flow will continue between all portions of the expanding population, but gradually localized subpopulations will develop in the most suitable habitats. When these subpopulations become partially isolated from one another, simply because the regions between suitable habitats are ecologically incapable of supporting snail populations for any considerable period, fragmentation has occurred. Each of the partially isolated demes may still retain the same genetic composition as the ancestral population, but the ecologic barriers to genetic exchanges between demes lay the groundwork for genetic divergence between them. Isolation, even if only spatial and partial, now becomes a decisive factor in evolutionary divergence. Without isolation the gene flow within the population precludes divergence, since no localized genetic changes may persist under the constant impact of recombination. Under these circumstances only sequential evolution takes place. With isolation, microevolution within the demes may produce markedly divergent populations. Isolation is therefore the key factor in the origin of new populations. Without it speciation is impossible.

To continue with the example of the insular snail, the possible results following initial fragmentation begin to relate to observed populational variation patterns. Let us assume that the entire island is now occupied by a series of demes, all of the same genetic composition and all partially isolated from one another. What forces will now operate to produce genetic differences between the demes? What are the determinants that will produce a pattern of random or nonrandom interspecific variation? And what roles will the conflicting forces of variation, drift, and selection play in producing evolutionary changes?

The fragmentation of the original population into a homogeneous series of demes is consistent with the known behavior of all living organisms. We have previously demonstrated that not all parts of an organism's environment are habitable and that demes usually become established only in small, suitable portions of a particular species range. In addition, with the exception of a very few creatures — certain of the oceanic plankton, for example — no large homogeneous populations are ever found under natural conditions, because the population invariably tends to break up into demic units. Our example of a population composed of demes having the same genetic characteristics is, for reasons given below, theoretical; if such a population did occur, it would not remain homogeneous for any length of time. From the very earliest stages of demic fragmentation each segment of the population will be under slightly differing pressures of mutation, selection, and drift. If the fragment deme occurs in a small but suitable portion of the habitat, the primary evolutionary forces will be modifying it from the moment of its origin. Selection acting on the variation within the deme may produce a trend toward higher levels of adaptive efficiency; or, if the population is small and selective pressure low, drift may fix striking nonadaptive features. Because the demes are partially isolated from one another, each may develop under its own particular combination of variation, selection, and drift, without great effect on adjacent demes or without being affected by them. Each deme proceeds along its own way by sequential evolution, virtually independent of other demes in the population. The principal bond between the demes remains occasional genic flow, which contributes to variation. Over a period of time each deme diverges from the others as a response to the impact of the evolutionary forces, but all the demes are still bound together by a web of occasional genic flow. As long as genetic exchange is possible, the demes retain fundamental similarities in common.

The general process described above may function to produce either of the two intraspecific variation patterns discussed in the previous chapter. If the impact of selection on variation is different in each deme's habitat or if selection pressure is low and the populations are small, random populational variation results. The example of the klipfish (*Gibbonsia*) illustrates this point (see Chapter 8). Each population within the species exhibits its own characteristic of body depth to body length ratio. The microevolution of this feature apparently occurs independently in each deme. The random variation pattern has been produced by a differential impact of selection and drift on the genetic components of each deme. Frequently, selection pressure is moderately strong for certain adaptive genotype combinations in ecologically different portions of a species range. Where this is the case, similar selection pressures may affect a large number of demes to the point that all of them will demonstrate the same adaptation. Nonrandom populational variation is the result, as indicated by race formation in the western rattlesnake (see Figure 8-2) and by cline formation in the Pacific herring (see Figure 8-5). In these examples, large segments of the species population, including many demes, have become similar in basic adaptive features under the impact of natural selection. Each deme within these species remains partially isolated, but demes under similar ecologic circumstances have attained a basic genetic similarity through selective pressures. The variation is nonrandom because it is directed by selection. Different segments of the population are adapted to differing environmental conditions within the species range.

Race formation follows two different pathways, depending upon historical factors. Many races and all clines are developed by changes occurring between segments of the species range continuously connected by modest but regular gene flow. In the example of the western rattlesnake, the Great Basin race (*lutosus*) and the populations adjacent to it (*oreganus*) probably developed from a once homogeneous population. Differentiation occurred as a response to the effects of selection pressures resulting from different environments. Nevertheless, contact through genetically intermediate demes (intergrading populations) is currently maintained, and differentiation appears to have proceeded, even though the developing races were bound together by genetic flow. The fact that significant changes were produced under these circumstances implies that selective pressures on the two portions of the original population must have been extreme. The two races are distinguished primarily by the obvious adaptive feature of coloration. The Great Basin form is a light-buff or drab-colored snake with small, dorsal, dark blotches and obscure lateral blotches. Its more widespread ally (*oreganus*) is a dark-gray, dark-olive, dark-brown, or black reptile with large, dark, dorsal blotches and conspicuous lateral blotches. (Some montane individuals are almost completely black.) These color-pattern characteristics appear to be adaptive in nature, and seem to correlate with soil color and composition in the two regions inhabited by these races. In the zone of intergradation, intermediate populations, transitional in their color-pattern characteristics, occur on soils of intermediate color.

A second means of race formation involves the same principle of genetic change in a segment of the species population. In this situation, however, the populations undergo genetic differentiation during a period when fragments of an ancestral population were completely isolated from one another in space. Allopatric race formation of this general type is relatively common and may be demonstrated by an example in which the populations currently retain complete spatial isolation. The spadefoot toad (*Scaphiopus holbrookii*) is usually regarded as a population made up of two races. One of these occurs in eastern North America from the Mississippi River eastward along the Gulf Coast-Atlantic Plain to Massachusetts, and northward to the Ohio River Valley. A second race occurs to the west of the Mississippi in Louisiana, east Texas, Oklahoma, and western Arkansas. The two populations are allopatric and were apparently separated from one another by the formation of the Mississippi River bottom lands. Evolution has proceeded along independent lines in each of the separated segments of the population, so that at present they differ markedly in skull characters and correlated head proportions. Laboratory experiments demonstrate that the two forms are completely compatible in genetic makeup. Crosses between the populations produce viable and fertile hybrids. Isolation in this example is purely spatial, and although differences in selection on the two allopatric populations have made them differentiate into two races, they still retain the potential for complete genetic exchange.

On occasion, races that originally developed allopatrically come into spatial association with one another after they have differentiated. Frequently they have diverged to a point where some reproductive isolation or a reduction in genetic-flow potential has been established. Such races intergrade only to a limited extent, and many of the demes of one or the other race along the boundary of contact are apparently uncontaminated by gene flow from the other race. A classic example of secondary contact and intergradation is provided by the long-nosed snake (*Rhinocheilus lecontei*). Two races in the southwestern United States are involved in this problem. One race is found in the arid deserts of California, Nevada, and Arizona; a second is found primarily around the desert periphery in semiarid habitats (Figure 9-1). The races differ markedly in scutellation and coloration. However, along the margins of contact between them a curious mixing of demes occurs. In this area some demes are typical of the desert race, others are typical of the desert-periphery form, and still others exhibit mixing of genetic materials from both types. Occasional demes of the desert race are found well within the range of the desert-periphery form, completely surrounded by demes of the latter. The reverse condition also occurs. The two races appear to have evolved allopatrically, differentiated to a considerable extent, and come back into contact with one another. Interbreeding is restricted as a result, but a moderate amount of genetic exchange takes place between the races.

Deme and race formation follow the same basic pattern. Before differentiation is possible, at least partial ecologic isolation must be established. When fragments of an original homogeneous ancestral population are isolated, variation, natural selection, and genetic drift operate differentially on the isolates to produce genetic divergence. Each population

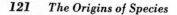

Long-nosed snake
(*Rhinocheilus lecontei*)

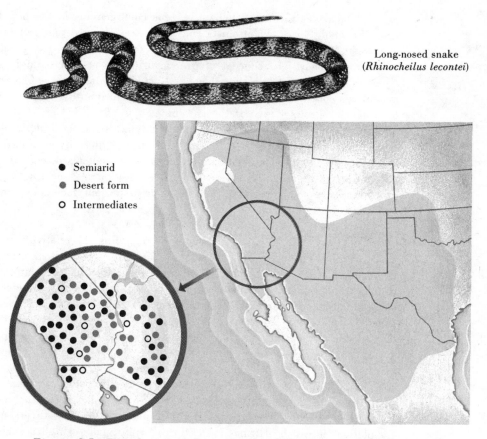

● Semiarid
● Desert form
○ Intermediates

Figure 9-1   Distribution of the semiarid and desert races of the long-nosed snake, Rhinocheilus lecontei.

fragment tends to develop its own variation pattern as a response to the peculiar combination of evolutionary forces operating on it. Races may be formed either as segments of a continuous population network undergoing different adaptive changes, or from completely allopatric portions of a previously connected ancestral population cut off from any genetic exchange by spatial isolation. Subsequently, a renewed contact may be established between the formerly isolated races, but there frequently tends to be some reproductive barrier between them, and genetic exchanges are restricted.

Among the many unsolved problems associated with speciation are the relative roles played by natural selection and genetic drift in populational divergence, especially at the demic and racial levels. Both these evolutionary forces tend to operate in the direction of reduced populational variability, but selection encourages adaptive genotypes, while drift may produce fixed neutral and nonadaptive gene combinations in the population. For many years evolutionists and geneticists assumed that gene frequencies within a population changed only in the direction of greater adaptive efficiency. The concept of genetic drift was at first strongly resisted as being statistically possible but biologically improbable. Experimental evidence now supports the fundamental principle that small populations of a species — even if the environment is homogeneous, selection and mutation rates the same, and initial genotypic frequencies identical — will become differentiated over a time period. The difficulty in applying this concept to natural populations is that the environments of each deme within species are slightly different, so that selection also operates. The problem remains to determine if drift is an important factor in demic and racial diversification.

Perhaps if we reexamine some of the variation patterns considered above, we may reach a tentative hypothesis regarding the relative significance of selection and drift in population formation. You will remember the example of the insular snails. At an early stage they had become fragmented into a number of isolated demes, each restricted to an ecologically suitable habitat. In our example these demes were originally identical in genetic composition. To simplify the model further, we will assume that they maintain identical mutation rates. Two possible influences may operate on the individual demes to produce diversity. If the several microhabitats occupied by each deme differ in ecologic characteristics, natural selection will operate differentially on each deme to produce gradual divergence. Even if such differences in the individual demic environments give rise to selection pressure, genetic drift may still produce divergence. Populations in which drift has been effective will usually exhibit a narrow range of variation in some nonadaptive feature that is fixed in the population. In addition, adjacent populations of snails will frequently exhibit striking differences where drift predominates. Since adjacent habitats of snail demes will usually be somewhat similar, there is a trend toward similarity in adaptive features among contiguous demes where natural selection is efficacious. Each demic population may then be characterized as a "selectee" or "drifter," depending upon the dominant force molding microevolution in the population. Reexamination of the examples discussed above will now give us some notion as to whether selectees or drifters are responsible for demic or racial formation.

In the case of the klipfish (*Gibbonsia*) cited as an example of random demic variation, it appeared likely that the particular character analyzed has been fixed not by selection but by drift. Each deme exhibits its own narrow range of variation in the ratio of body depth to body length, a characteristic not correlated with environmental differences. The general conclusion may be made that random demic variation is the result of a preponderance of drifters among the population units. Random demic variation is common in natural populations, and it may be concluded fairly easily that drifters are at full strength as components in species of this type.

Selection is the strongest element in most well-analyzed species exhibiting ecogeographic racial differentiation. In the western rattlesnake example, selectee populations apparently predominate over drifters, and contiguous demes under similar selective pressures come to form a network of demes with basic adaptations in common. This network of demes is a race.

Not all situations are as clear-cut as the two examples above. In the cases of the spadefoot toads and the long-nosed snake, it seems likely that the observed variation pattern is due to selection. However, many examples exist where small allopatric segments of a species population may be drifters, and other demes selectees. An instance of this type is provided by the brown-shouldered lizard (*Uta stansburiana*). This reptile has a wide range in western North America and forms thousands of demes. In the Gulf of California region, isolated allopatric populations occur on a number of islands. Mainland lizards are usually brownish, with dark, dorsal markings and numerous light-blue and bright-orange or yellow flecks. One insular population on Isla San Pedro Martir in the northern Gulf is typified by all individuals being a uniform slate gray. Significantly, this island is hardly more than a small pile of boulders emergent from the surrounding sea, but the color of the rocks is precisely the same gray hue as that characterizing the lizards. This small population has been under intensive predation by birds, and natural selection obviously has molded the genotypes in the direction of perfecting protective coloration. Another population on Isla Tortuga is typically represented by individuals that are dorsally a uniform black. This island is a volcanic cone covered by lava flows and volcanic rocks. Again selection has favored the acquisition of a coloration that resembles the black lava backgrounds on which the lizards live.

Elsewhere in the Gulf, drifter populations of this animal have been established. On Isla San Pedro Nolasco, which is composed of light-gray and whitish boulders, the lizards of this species are a bright copper green. They are extremely obvious on the light rocks and their coloration is non-adaptive to a high degree. Apparently drift has been the important influence here. Significantly, on Isla Santa Catalina, some distance to the southwest, another population of bright-green lizards occurs. This island is a light-colored granitic ridge, and the lizards are again obvious and subject to easy predation by birds. Drift seems to be the only possible explanation of this situation, and since the Catalina population is rather distinct from the San Pedro form in other features, random fixation of the green coloration seems to have occurred on two different occasions in two different situations.

The conclusion to be drawn from this example is clear. In many species some populations are drifters and others selectees. Still others, we may assume, may be equally affected by the conflicting forces of drift and selection and might be called "neutralists" in the constant struggle for supremacy. Probably most species contain some demes of each of the three types. All three situations contribute to demic fragmentation and diversification, but in varying amounts, depending upon the complex of conditions under which the species lives.

THE ORIGIN OF REPRODUCTIVE ISOLATION You will recall from the discussion in the preceding chapter that speciation produces two distinctive patterns of isolation between related species. In some cases the forms are allopatric and are prevented from interbreeding by spatial isolation. In others the species are sympatric and are reproductively isolated by one or more isolating mechanisms. Theoretically, these patterns may have been produced in either of two ways; by *allopatric speciation* or by *sympatric speciation* (Figure 9-2).

In allopatric speciation, evolution may develop in two spatially isolated populations descended from a common ancestor. Differences in the interaction of variation, selection, and drift will operate in these frag-

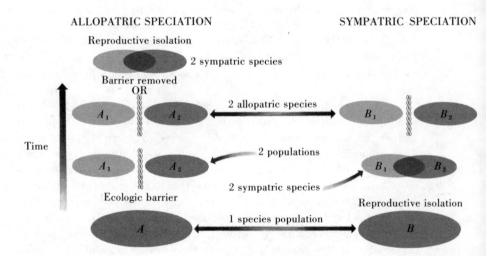

Figure 9-2 Diagrammatic representation of the process of allopatric and sympatric speciation.

ments in exactly the same process of microevolution described for deme and race formation. If the populations remain separated for a long enough time and if the interacting forces of evolution — particularly selection — operate to produce divergence, allopatric species result. When an ecologic barrier separating allopatric populations is removed, the populations may become sympatric and, if reproductive isolation has been developed during the period of allopatry, may come to share a portion of one another's range (see Chapter 8). Allopatric speciation follows microevolutionary pathways and largely depends upon spatial isolation, time, and differences in environment, as expressed in selection, to produce new gene combinations.

Sympatric speciation may produce similar results. If reproductive isolation between segments of the same population were to arise instantaneously and directly, two reproductively isolated populations would result. If one or both populations moved out of the original habitat, the species would be allopatric. If the two populations remained in the same area, sympatric species would be recognized (Figure 9-3). Once repro-

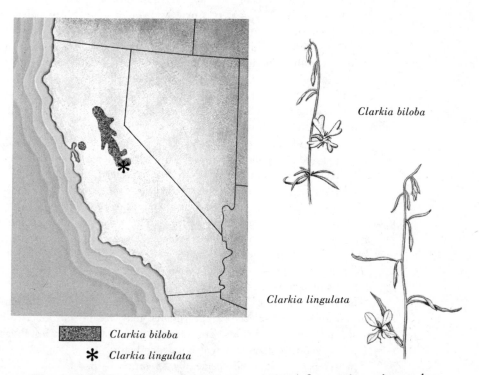

Clarkia biloba

Clarkia lingulata

▨ *Clarkia biloba*

✳ *Clarkia lingulata*

Figure 9-3 Distribution of sympatric species of the evening primrose family, illustrating sympatric speciation: Clarkia biloba (N = 8) *apparently gave rise to the form* Clarkia lingulata (N = 9). *Although the two species occur sympatrically, no interbreeding or gene exchange takes place.*

ductive isolation is established, each population follows its own evolutionary course.

Allopatric speciation has long been recognized as a primary method of species origin. All evidence from the manner of deme and race formation and from the frequent intermediate situations where originally allopatric populations, not yet fully isolated reproductively, have become sympatric (see Figure 9-1) substantiates the concept of allopatric speciation. Somewhat more controversial is the theory of sympatric speciation. Originally discounted as a primary factor in evolution, it has now become recognized as a powerful influence in forming species of higher plants. Allopatric speciation has been adequately covered in the section on deme and race formation, but further comment is required regarding the sympatric process.

Sympatric speciation may occur, at least among higher plants, in four principal ways: (1) by polyploidy, (2) by hybridization between distinct species, (3) by self-fertilization, and (4) by apomixis. *Polyploidy* is a spontaneous increase in chromosome numbers that usually results in a new individual with $3N$, $4N$, $6N$, or $8N$ chromosomes. Since polyploidy is usually caused by failure of proper meiosis, one set of seeds may produce a number of individuals with a similar polyploidy. Polyploid individuals are usually genetically isolated from the parental species. In the early years of divergence, the diploid and polyploid species may occur together and may be practically indistinguishable. However, once reproductive isolation is established, differences in gene combinations will result through microevolutionary forces.

Because of the manner in which pollen is distributed, chance fertilization of one species by pollen from a closely related form is not uncommon in plants. Hybrid individuals frequently are produced. Sometimes the hybrids and their genes are reintroduced into the gene pools of one or both parental species by backcrossing to the parental types. Frequently the hybrids have faulty meiosis and produce hybrid polyploid offspring reproductively isolated from all other species. These polyploid species are now known to be very common. In the tarweeds (*Madia*), for example, the species *M. citrigracilis* ($N = 24$) apparently was produced through hybridization between *M. citrodora* ($N = 8$) and *M. gracilis* ($N = 16$).

The development of the habit of self-fertilization, an evolutionary shift by no means restricted to higher plants, may create reproductive isolation instantaneously. Unless cross-fertilization becomes reestablished later, each descendent stock will evolve independently.

The sympatric origin of species by *apomixis* is also common among plants and many lower animals. Apomixis is any asexual reproductive process that replaces the sexual method. A variety of apomictic reproductive types are known, but all produce offspring from unfertilized gametes or by vegetative means. When apomixis arises suddenly in an originally sexual population, the asexually reproducing individuals immediately form isolated gene pools. Genetic isolation is established and once again the descendent stocks evolve independently.

Unfortunately, what constitutes a gene pool or species in an asexual organism is a very real biological question. Groups of rigid self-fertilizers or apomictic reproducers are almost impossible to treat as species, since the offspring of each individual are genetically identical to the parent and reproductively isolated from all other individuals. Evolution in such forms is influenced much more strongly by mutation and drift than by selection, because all offspring of a particular parent will have the same genotypic makeup as the parent. For convenience of discussion and since most organisms cross-fertilize and reproduce sexually, no further reference to restricted self-fertilizing and apomictic forms will be made in this book. The subject is a fascinating one and is in need of a great deal of study.

The one essential problem in speciation, namely, the manner in which isolating mechanisms become established, remains unsolved. Evolutionists have proposed a number of solutions, but no single concept appears to be equally applicable to all cases, and most of the suggestions have not been rigorously tested. The mechanism of the origin of reproductive isolation between populations produced by sympatric speciation is relatively clear. Either the chromosome number or composition is incompatible for interbreeding (polyploidy and hybridization), or interbreeding ceases (self-fertilization and apomixis).

Mechanisms suggested for the origin of reproductive isolation in allopatric speciation include the following.

1. If two populations are ecologically isolated for a long enough time, differential mutation, drift, and selection will ultimately lead to gene combinations producing reproductive isolation.

2. If two formerly ecologically isolated populations become sympatric, selection will operate against any hybrids produced by accidental interbreeding and will favor and reinforce all reproductive isolating mechanisms.

3. If the genetic controls of hybrid sterility are correlated with genetic features with a positive selection value, then genetic incompatibility will result from any crosses.

4. If sterility-producing genes are neutral or even nonadaptive, they may become fixed by genetic drift in small populations.

The question still remains of how reproductive isolating mechanisms arise in allopatric populations and prevent interbreeding when the populations become sympatric. Solution of this problem is one of the challenges in the study of modern biology.

SPECIATION AND Speciation may occur at a number of evolu-
ADAPTIVE RADIATION tionary levels. Divergent fragmentation of the
species through ecologic isolation leads to
microevolutionary change in each populational segment. New species may
also arise through microevolutionary shifts in allopatric populations or by
sympatric speciation followed by microevolution. Speciation may also
involve a number of additional factors.

A classic example of speciation involving the interplay of complex
forces leading to adaptive radiation at the species level is provided by
Darwin's finches (*Geospiza*) of the Islas Encantadas (Galápagos Islands).
Significantly, the study of these birds in their native habitat gave Darwin
a major insight into evolutionary processes. The islands are located astride
the equator, 500 miles west of the coast of South America (Figure 9-4).
The several closely related genera of finches found on the islands appear
to have been derived from a single common ancestor. The group is distinctive and is not represented on the mainland.

The Galápagos are of volcanic origin and at their lower elevations
are covered by scrub, thorny brushes, trees, and cacti. On most of the
islands the upland areas are covered by humid woods. The ground finch
(*Geospiza*) is represented in the islands by six species. Some members of
the group are seed eaters and nest in a variety of low plants, others feed
on the cactus fruits and nest in the cactus. Some species occur in the dry
lowlands, others in the humid uplands. In the lowlands on the large central
islands three species occur, a large, a medium, and a small ground finch.
A single species of cactus finch is also present. On the smaller peripheral
islands the pattern is very different (Table 9-1). In a series of fragmented
islands, allopatric speciation is probably responsible for evolution, particularly since these finches are very weak fliers. Moreover, several of the
insular populations of particular species differ from one another in essential characteristics. But the most significant feature of the distribution is
that it shows that speciation has involved more than microevolution. What
are the peculiarities of adaptive radiation in this example?

First of all, note that where there is no ecologic competition, some
species occupy several different habitats and fill several ecologic roles
(*G. difficilis* on Culpepper). Where competition is present, the species have
different ecologic relations on different islands (*G. difficilis* on Wenman
and Culpepper as compared with the same species on Tower or the central
islands). Competition between sympatric populations has a significant
effect in molding radiation. No competition favors generalized habits;
marked competition leads to specialized ecologic roles.

A second force in radiation is ecologic accessibility. *G. difficilis* is
blocked out from the small ground finch and cactus finch roles on the
central islands (very likely it was forced out of them by other species
competition), but where these ecologic situations are open, as on Wenman,

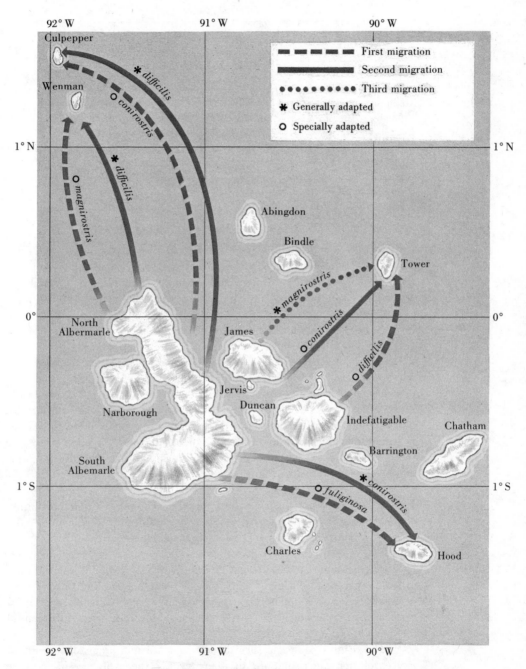

Figure 9-4 The Islas Encantadas (Galápagos Islands) and speciation in Darwin's finches. This example demonstrates the tendency for general adaptation to be replaced over time by special adaptation. (See text for explanation.)

Table 9-1   *Darwin's Finches*

ECOLOGIC ROLE	*Central*	*Tower*	ISLANDS *Wenman*	*Culpepper*	*Hood*
Large ground finch	*magnirostris*	*magnirostris*	*magnirostris*	*conirostris*	*conirostris*
Medium ground finch	*fortis*				
Cactus ground finch	*scandens*	*conirostris*	*difficilis*	*difficilis*	*conirostris*
Small ground finch	*fuliginosa*	*difficilis*	*difficilis*	*difficilis*	*fuliginosa*
Humid ground finch	*difficilis*	. . . . no humid woods			

difficilis moved into them. Correlated with availability in radiation is the multiple invasion of available adaptive zones. The cactus finch role has been assumed by three different species in the islands, and the multiple probing of divergent lines into similar adaptive relations is always typical of radiating evolution.

A final feature of adaptive macroevolution at the species level is provided by this example. Every pattern of adaptive radiation so far studied indicates a sequence of change involving the replacement of general adaptation by special adaptation through time. When one species fails to compete successfully with another closely related form, it is usually because the successful species retains a greater residue of general adaptation. Usually when a population that is generally better adapted arrives in a new environment, it takes over the major portion of available ecologic roles, while the old resident population must become specially adapted. Examples are outlined in Figure 9-4.

It is seen from the discussion in this chapter that microevolution and allopatric speciation at its simplest are essentially identical. Populational divergence is dependent upon reproductive isolation, although this isolation between demes and races may be temporary. New reproductively isolated species may arise through the interaction of ecologic isolation and differential microevolution. The origins of isolating mechanisms between originally allopatric populations remain a mystery. Sympatric speciation rests on the establishment of reproductive isolation followed by differential microevolution.

The origins of species are diverse and are not simply the result of microevolution and isolation. In many groups, the evolving species take on new adaptive relations of a high order, involving major shifts in ecologic roles. Among the principal features of adaptive radiation in speciation are the interactions of interspecies competition, ecologic accessibility, multiple attempts at taking on the same new ecologic role by different populations, and the significant interplay of general and special adaptation. Evolutionary divergence at every level is a reflection of constant experimentation in innumerable directions and along a broad boundary of new organism-environment relations; its movement is driven by the elementary evolutionary forces, intensified by isolation and modified by the characteristic features of adaptive radiation.

FURTHER READING

Grant, V., *The Origin of Adaptations*. New York: Columbia University Press, 1963.

————, *Plant Speciation*. New York: Columbia University Press, 1971.

Mayr, E., *Populations, Species and Evolution*. Cambridge: Belknap Press, 1970.

Stebbins, G. L., *Variation and Evolution in Plants*. New York: Columbia University Press, 1950.

Wright, S., "Population structure and evolution," *Proceedings of the American Philosophical Society*, vol. 93 (1949), pp. 471–478.

chapter 10

*Evolution
above the
Species
Level*

Our previous discussions have focused on the forces and circumstances responsible for populational evolution. The effects of variation, drift, and selection on gene frequencies and gene combinations in sequential change or populational fragmentation have been our principal concern. Our attention is now directed toward a consideration of the panorama of evolution, not in terms of the basic forces of populational change and divergence, but in terms of an understanding of the evolution of major groups of organisms.

The essential features of microevolution and speciation are now fairly well understood by biologists, but the complex processes leading to evolution on a grander scale remain an area inviting investigation. At the present time, we have only the most shadowy impressions of the forces contributing to the adaptive radiation and diversification of life. For example, can the evolution and diversity of the flowering plants be explained simply on the basis of microevolutionary change, or are other forces contributing to macro- and megaevolution? The interaction of variation, selection, and drift and the taking on of new adaptive efficiency must play an exceedingly important part in these processes, but is the grand pattern of evolution the result only of simple population change? To many paleontologists and to those biologists interested in major evolutionary shifts, the question remains open. No satisfactory mechanism or mechanisms have been proposed that might explain these phenomena, but the characteristics, modes, patterns, and pathways of evolution at this level all suggest that other factors besides those operating at the population level must contribute to adaptive radiation and to the origin of new biological systems. The dim outlines of the course and general features of evolution above the species level are the subjects of this chapter. The history of major changes in life is well documented, in many cases by the fossil record, and glimpses of common patterns of evolution are provided by a comparison between flowering plants and reptiles, corals and ferns, lungfish and horses. Nevertheless, the elucidation of the extremely complicated processes contributing to the development of major groups and new major general adaptations remains a challenge to all biologists.

THE EVOLUTION OF ADAPTATION   George Gaylord Simpson, for many years at New York's American Museum of Natural History and at Harvard University, and one of the principal paleontological contributors to the synthetic theory of evolution, has developed a neat conceptual framework for describing major evolutionary patterns. We already know that at any moment in time the interaction of organisms and their environments define a series of broad or narrow adaptive fields or zones. All members of the same major group (for example the crustaceans) share one major adaptive zone because of their common possession of a complex of general adaptations. Within the broad zone, each species of crustacean occupies a distinct but narrow field of its own because of the species' peculiar combination of special and general adaptations. Each kind of organism is an adaptive type discontinuous from other adaptive types; a crab is distinct from a barnacle (both are crustaceans), a snake, from a man or a sunflower. Simpson suggests that for purposes of discussion the adaptive types or zones may be represented diagrammatically as bands or pathways on an adaptive grid (Figure 10-1). The discontinuities between the major zones (A, B, C, D) are ecologic discontinuities or unstable ecologic zones. The adaptive zone itself is an ecologic role or characteristic relationship be-

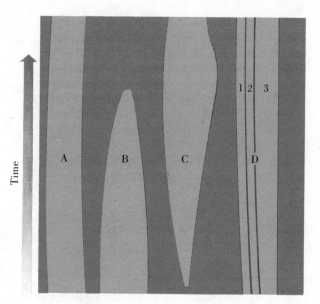

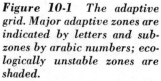

Figure 10-1 The adaptive grid. Major adaptive zones are indicated by letters and sub-zones by arabic numbers; ecologically unstable zones are shaded.

tween organism and environment. Although the actual grid is of course very complex, with many subzones (1, 2, 3), a simplified form is satisfactory for our discussion. The diagram is extremely useful in attempting to describe adaptive evolutionary change.

Figure 10-2 An adaptive grid diagram of the evolution of terrestrial plants, indicating major breakthroughs and invasions of new adaptive zones.

The evolution of adaptation may be summarized in terms of the grid as follows: Changes in adaptation involve movement of evolutionary lines within subdivisions of the subzones (microevolution), either from one major zone or subzone into others (macroevolution), or from one major set of zones into another (megaevolution).

The basic feature of evolution above the species level is the movement of a group of organisms into a new adaptive zone. In order to move across the zone of ecologic instability (an environment of strong negative selection) into the new adaptive zone, an organism must have evolutionary and ecologic access to it. In other words, the group must already have some characteristics adaptive to the new zone, and the zone must be unoccupied by a strong competitor. An example is provided by the evolution of terrestrial plants (Figure 10-2).

Another case in point, which illustrates the concept of the adaptive grid in more detail, is the evolution of the lungless salamanders (Plethodontidae) (Figure 10-3). This family is the most successful and numerous group of living salamanders, with representatives in Europe, North Amer-

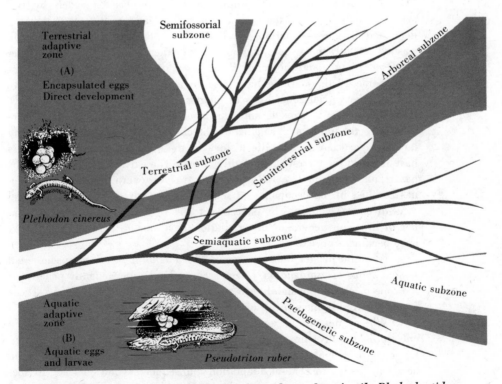

Figure 10-3   *Evolution in the lungless salamanders, family Plethodontidae, expressed on an adaptive grid diagram.*

ica, and tropical America. The ancestral adaptive zone for the group can be characterized as a mode of life in semiaquatic montane situations. The adults of the ancestral stock probably occurred in and around the margins of small mountain streams. Their life history reflected this situation, since they apparently laid delicate eggs in the water which hatched into a free-living aquatic larva with functional gills. After undergoing further development, the larva transformed into a semiaquatic salamander. A number of living genera (*Desmognathus, Eurycea, Gyrinophilus, Pseudotriton,* and *Typhlotriton*) in the group retain these habits and continue to occupy the ancestral semiaquatic adaptive subzone. Several descendent evolutionary lines have become completely aquatic in habits. In one genus (*Leurognathus*), closely allied to *Desmognathus* in the semiaquatic group, the adult is strictly aquatic and hides under rocks on the bottom of fast-moving mountain streams. A second group of completely aquatic forms occupy the paedogenetic subzone; they never transform, but become sexually mature and breed while still in a gilled larval state. Some species of *Eurycea* and *Gyrinophilus* have evolved by this means, while two distinctive genera allied to the former also live in the paedogenetic subzone.

A fourth subzone, the semiterrestrial, has been invaded by two species. In these forms the eggs are unspecialized aquatic types that are deposited in moist situations on land. When the larvae hatch they are washed into streams, where they undergo further development. One species of *Desmognathus* and the genus *Hemidactylium* occupy this subzone.

Some Plethodontidae have also invaded a second major adaptive zone, the terrestrial zone. Eggs of lungless salamanders in this zone have a special protective capsule and are deposited on land. Development proceeds directly and a small salamander hatches out of the egg. No free-living larval stage occurs in the life history of these forms. These salamanders have invaded the terrestrial zone three times. The earliest of these invasions has resulted in adaptive radiation into twelve genera. Further evolution has resulted in the occupancy of semifossorial (burrowing) and arboreal (tree-dwelling) subzones by terrestrial forms or their descendants. A number of different genera of this radiation have one to several species that have become arboreal.

At a much later date one genus (*Phaeognathus*) and a species of a related genus (*Desmognathus*) each seem to have independently penetrated the zone of ecologic instability to reach the terrestrial adaptive zone. These forms apparently have incapsulated eggs with direct development.

When a new major zone or group of zones is occupied, the first zone entered is usually the widest and requires the least special adaptation. Later, more special zones will be occupied. The initial breakthrough to the new zone is due to general adaptation. Subsequent evolution leads to specialization. This pattern also fits the macroevolutionary speciation of Darwin's finches, discussed in the previous chapter, and it is suggested that you fit that example to the grid concept. In the example of the lungless salamanders, can you identify some special adaptations leading to invasion of subzones and the key new adaptation allowing for occupancy of a new adaptive zone? We will return later to the grid as an aid in the explanation of other features of evolution above the species level.

EVOLUTIONARY   Among the most significant features of macro-
TRENDS   evolution is a progressive, sustained tendency
for certain characters to develop along an evo-
lutionary line. Trends of this sort are numerous in the fossil record. Long-term continuing trends rarely appear in only one structure, but almost always involve a complex of different features. Trends are, of course, produced by the driving force of natural selection operating within the limits of a particular adaptive zone or subzone. Evolution is not random, although certain elements in the process are random, and trends leading to greater adaptive efficiency are to be expected. Evolutionary trends are generally adaptive movements along one pathway, but they are never exclusively sequential and always involve divergent and repeated taking on of one or the other characters important in the trend. The populations undergoing change are constantly experimenting within the adaptive pathway, and parallel probings of a new subzone by related but different lines are the rule. In a sense, to refer to these patterns of divergent and multiple exploration of adaptive relations as trends is misleading, because a straight-line evolution of characters is not involved.

The classic example of evolutionary trends in macroevolution is provided by the horse family, Equidae. A summary of the evolutionary history of the family is presented in Figure 10-4. It will be seen at once that evolution from the small Eocene ancestor *Hyracotherium* to modern horses (*Equus*) was not along a straight line. Actually, the Eocene genus is almost as much a rhinoceros as it is a horse. The diagram is obviously greatly oversimplified, but it demonstrates very nicely the gradual change from a doglike, browsing creature with padded feet, to a horse with grazing habits and hoofed springing feet. This evolutionary history reaffirms the pattern of multiple divergences leading to special adaptation.

Natural selection provides the principal nonrandom force to evolutionary change, and apparent trends are due to the creative directive force of selection. Once a new adaptive zone is occupied, only a limited number of possibilities are open to the evolving stock. Each succeeding special adaptation limits the possibilities for future evolution. Attempts to fit the basic divergent pattern of evolutionary change into a rigid sequential

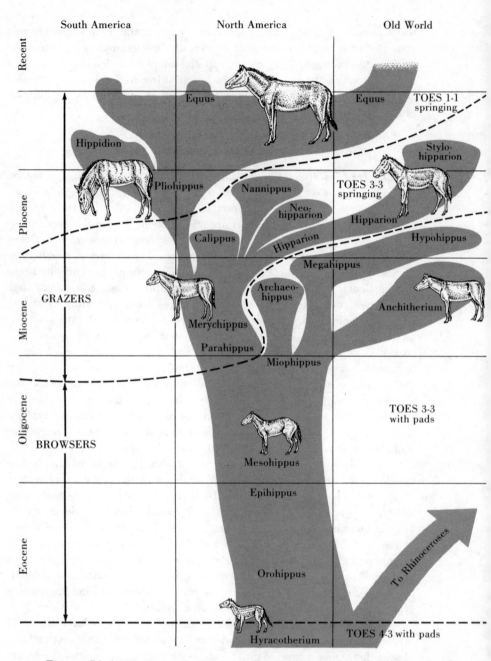

Figure 10-4 *Evolution in horses: from the ancestral browsing types with four toes on the front feet and three on the hind feet, padded as in dogs, to modern grazing forms with only one toe on both front and hind feet, hoofed and modified for springing action in running.*

scheme falsify the record and lead to erroneous conclusions. Trends in evolution exist, but they always involve multiple diverging trials within the adaptive zone.

ADAPTIVE
RADIATION IN EVOLUTION
Macroevolution, or adaptive radiation, is a significant, observable feature of change in all groups of organisms. Radiation as a factor in speciation has been considered in an earlier chapter and need not concern us here. Adaptive radiation above the species level is of greatest interest, since the facts it supplies must form the basis for understanding the processes responsible for the grand pattern of evolutionary diversity. The evolution of the reptiles, one of the best-documented histories in the fossil record, will serve as a fine example of macroevolution and as a point of reference for its analysis.

The class Reptilia first appears in the fossil record in Pennsylvanian times (310 million years ago). Adaptive radiation in the group occurred between Permian and Cretaceous times, and living reptiles are derived from Cretaceous ancestors. The initial success of the reptiles stems from a megaevolutionary shift from aquatic to completely terrestrial development; reptile eggs, like bird eggs, do not need to be immersed in water to survive. The basic stock of reptiles were the Cotylosauria (stem reptiles), characterized by a primitive skull (Figure 10.5). [The Mesosauria (Pennsylvanian-Permian) also belong here.] All other groups of reptiles are derived from this generally adapted group. Because of the major differences in skull type (each as a modification to ensure more efficient jaw musculature) and other changes, the reptiles are classified into six major subclasses (Figure 10-6). Within each group adaptive radiation has occurred, producing a magnificent diversity of reptile species.

The basal stock of anapsid reptiles included the generalized order Cotylosauria (Pennsylvanian-Triassic) and the extremely specialized armored turtles, order Testudinata (Permian-Recent).

Derived from the anapsids was the ancient synapsid line, with three principal radiating groups. The oldest are the fin-backed pelycosaurs, order Pelycosauria (Pennsylvania-Permian), and their less specialized relatives, the mammallike reptiles, order Therapsida. From the latter order developed man's ancestors, the mammals, sometime in the Cretaceous.

Also appearing early in reptile history were the aquatic members of the subclass Euryapsida, including the fishlike Ichthyosauria (Triassic-Cretaceous) and the primitive order Protosauria (Permian-Jurrasic), which gave rise to the giant swimming pleisosaurs, order Sauropterygia (Triassic-Cretaceous).

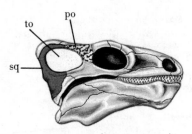

Synapsid (Edaphosaurus)

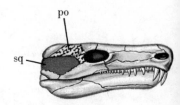

Anapsid (Limnoscelis)

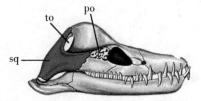

Euryapsid (Plesiosaurus)

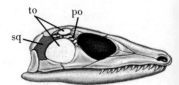

Diapsid (Youngina)

sq=Squamosal
po=Postorbital
to=Temporal opening

Figure 10-5 Basic skull types in reptiles.

The two largest groups involved in reptilian radiation are both characterized by a diapsid skull. The more primitive subclass, Lepidosauria, is represented by three orders, the basal Eosuchia (Permian-Triassic), the order Rhynchocephalia (Triassic-Recent), and the lizards and snakes, order Squamata (Jurassic-Recent). The tuatara of New Zealand is the lone surviving rhynchocephalian.

The second diapsid group, the subclass Archosauria, or ruling reptiles, is descended from the lepidosaurian order Eosuchia. The basal archosaurian stock comprises the order Thecodontia (Permian-Triassic) and from it arose five divergent specialized orders of reptiles that came to dominate the terrestrial environment during Mesozoic times. Apparently the thecodonts represent a breakthrough into a new series of adaptive zones by the taking on of a new general adaptation; four of the five orders of higher archosaurians differentiated through special adaptation; the fifth in its turn represents another major evolutionary shift founded on a new general adaptive type. Two of the thecodont derivatives, order Saurischia (Triassic-Cretaceous) and order Ornithischia (Triassic-Cretaceous), were enor-

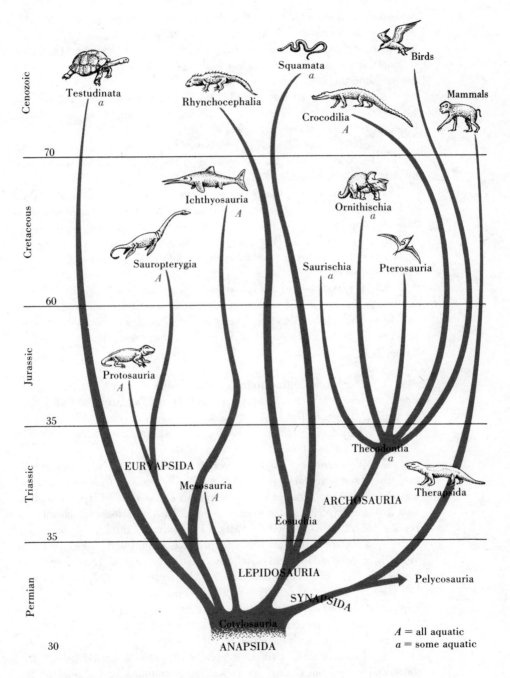

Figure 10-6 *General pattern of reptile evolution.*

mously successful, with many families, hundreds of genera, and innumerable species. Members of the two groups, frequently called "dinosaurs," included such diverse forms as small lizardlike reptiles, gigantic monsters, delicate bipedal species resembling flightless birds, and successful aquatic forms. None survived into the Cenozoic.

A third divergent line included the flying reptiles, order Pterosauria (Jurassic-Cretaceous). Gliders were probably the ancestral line of this order, but later forms were capable of flight.

The fourth line of special adaptation forms the aquatic order Crocodilia (Triassic-Recent), very successful in the past but today represented by only twenty-one species.

The fifth and final thecodont stock also took up the habit of gliding and then flying, and this habit, coordinated with the development of a new general adaptive complex of other features, led to occupation of the adaptive zone of aerial life. This group—the feathered reptiles, or birds—has undergone its own extensive adaptive radiation to form the class Aves (Jurassic-Recent).

Certain features of adaptive radiation found in the evolution of reptiles are principal characteristics of macroevolution in all groups, from tree ferns and trilobites to seed plants and insects. These common elements provide the key to understanding the process of evolution above the species level.

1. All macroevolution follows the acquisition of new general adaptation or entrance into a new adaptive zone. Darwin's finches radiated after an apparently generalized finch ancestor arrived to occupy the previously unoccupied Galápagos Islands. Reptiles radiated after completely terrestrial development was established as a general adaptation.

2. Macroevolution always involves evolutionary divergence. Macroevolution is not linear but radiating. Usually radiation follows general adaptation and the invasion of a new adaptive zone through special adaptation in different divergent descendent lines. The radiation of archosaurian reptiles from the generalized thecodonts is a typical example.

3. Adaptive radiation tends to produce evolutionary lines that converge in special adaptation with other distantly related groups differing in their matrix of general adaptation. The ichthyosaurs show a marked evolutionary convergence with a number of fish groups in their body form, manner of locomotion, food habits (fishes), and free-swimming pelagic life. They are convergent in special adaptations but still share a group of general adaptations with all other reptiles.

4. Macroevolution produces groups of parallel special adaptations among divergent but related stocks sharing a common background of general adaptation. Among the reptiles, group after group has invaded adap-

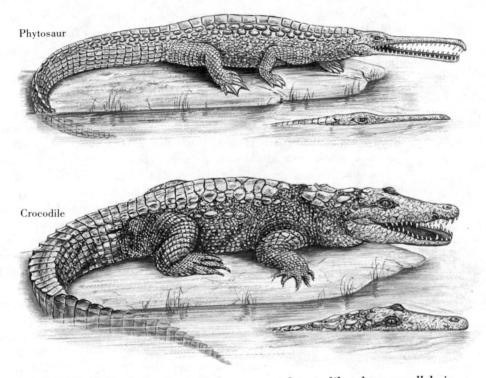

Figure 10-7   Comparison of phytosaurs and crocodiles shows parallels in evolution.

tive zones in aquatic habitats. Representatives of every order except the cotylosaurs, the synapsid orders, the eosuchians, rhynchocephalians, and flying reptiles have invaded the water and become completely aquatic. Repeated trials of the same general habitat or group of adaptive zones by divergent lines are always typical of adaptive radiation. A corollary of this principle is the feature of ecologic replacement in macroevolution. Where there are repeated evolutionary experiments with the same broad group of zones, some groups arise, flower, and become extinct, to be replaced by parallel groups that invade the same zone and undergo differentiation in their turn.

The phytosaurs (Triassic), of the order Thecodontia, and the crocodiles (order Crocodilia, Triassic-Recent) provide an example of both parallelism and ecologic replacement. The phytosaurs were very successfully adapted to aquatic life, but as they became extinct the crocodiles, derived from a different group of thecodonts, replaced them (Figure 10-7). The special aquatic adaptations shared by the two groups included location of eyes and nostrils dorsally, a large muscular tail used in locomotion, similar dentition for capture of aquatic prey, and devices for insuring respiration while nearly completely immersed in water. The specialized phytosaurs have been replaced by the specialized crocodiles and their allies, who in their turn are becoming extinct.

5. As a rule, macroevolution ultimately leads to extinction. As general adaptation is replaced by special adaptation, groups become rigidly specialized to narrow adaptive subzones and are unable to move into new major zones. Since all adaptive zones must finally change and disappear, all groups locked into a narrow zone are doomed. Evolutionary progress consists of moving out of old zones into new ones through the acquisition of new complexes of general adaptations. The fossil record composed of the bones of extinct organisms not only forms the material for study by evolutionists, but at the same time offers immutable evidence of extinction as the ultimate fate of every line. Two reptile groups may be used to illustrate the reality of extinction. The order Rhynchocephalia was extremely successful up through the Cretaceous but is unknown as fossils in the last 75 million years. A single species, the tuatura, is a relict of the order and survives today in New Zealand. It is sometimes called "a living fossil," although the term is a semantic absurdity. Relicts of a similar type are known in other groups and seem to survive through the retention of a modicum of general adaptation or as extremely specialized forms hanging on to existence by the thinnest margin. The extinction of the two dinosaur orders emphasizes again that evolutionary success through special adaptation is ephemeral. The road to extinction is paved with the remains of beautifully but specially adapted types. Final evolutionary victory over the malevolent environment requires progressive movement into new broad adaptive zones. Many lines never move out of the old zones; others fail in the attempt; but those which succeed form the advancing army of organic diversity and increasing general adaptive efficiency.

A common denominator runs through the entire picture of macroevolution, whether seen in Darwin's finches or in the ruling reptiles. At every step in evolution, natural selection creates a vast array of adaptive experiments that occasionally break out of their present adaptive zones into new groups of zones. Every new breakthrough is dependent upon the breakthroughs previously achieved; an alga did not become a seed plant in a single step, nor did fish develop directly into man. The adaptive shifts directly responsible for sunflowers or human beings were not possible without all the progressive evolutionary shifts that preceded them. The individual shifts depend upon opportunity, ecologic access, and the taking on of new general adaptation. The reptiles could not have evolved if amphibians had not previously invaded the land; birds and mammals are new permutations dependent upon the successful general adaptation of reptiles.

The adaptive experiments created by selection within an adaptive zone are of two types. Many are special adaptations taking their possessors into narrower and narrower subzones. Others, and these are at the base of any breakthrough to a new zone, are in the direction of the outermost limits of the present zone. Repeated attempts and combinations are apparently developed by selection until one line makes the breakthrough. An example of the constant, divergent probing of the limits of an occupied zone by many different branches of the same line is provided by frogs and their many adaptations to reproduction out of the water (Figure 10-8). Frog evolution in these cases parallels what must have been happening among Pennsylvanian amphibians as selection favored some kind of escape from aquatic development. One of the experiments took advantage of the opportunity and led to the reptiles. No frog has ever made the shift to a completely new adaptive zone, but if the opportunity arises, one of the probing lines may move across the no-frog land of ecologic instability into a new unoccupied major zone. The multiple attempts at crossing the barrier usually lead to extinction or failure, but at the foundation of every successful breakthrough a series of divergent testings of the unstable environment has occurred.

THE ORIGIN OF NEW BIOLOGICAL SYSTEMS The central problem confronting all forms of life is the entrance into new major adaptive complexes of zones through development of new general adaptive types. Megaevolution, or the origin of new biological organizational plans, is rare and shares features common to microevolution and macroevolution. Only a few major general adaptive types have developed during the history of life, but almost all of them persist without extinction, although a few have perished and others are relict. All the phyla and most classes of microorganisms, plants, and animals represent marvelously complex and coordinated groups of general adaptations, each forming a basic, distinctive biological organization, unique and dominant in a broad range of characteristic adaptive zones. Fewer than 200 of these major biological organization plans are known to have developed in 3 billion years. The origins of these systems are the most important of all evolutionary events; at the same time, the processes leading to these events remain the least studied of biological problems.

Megaevolution, insofar as the process has been analyzed, has the following clear-cut characteristics.

1. The breakthrough always follows evolutionary experimentation

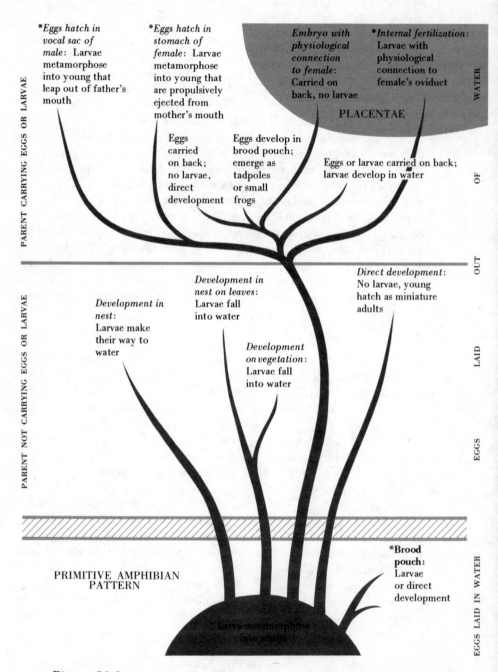

PARENT CARRYING EGGS OR LARVAE

*Eggs hatch in vocal sac of male: Larvae metamorphose into young that leap out of father's mouth

*Eggs hatch in stomach of female: Larvae metamorphose into young that are propulsively ejected from mother's mouth

Embryo with physiological connection to female: Carried on back, no larvae

*Internal fertilization: Larvae with physiological connection to female's oviduct

PLACENTAE

Eggs carried on back; no larvae, direct development

Eggs develop in brood pouch; emerge as tadpoles or small frogs

Eggs or larvae carried on back; larvae develop in water

PARENT NOT CARRYING EGGS OR LARVAE

Development in nest: Larvae make their way to water

Development in nest on leaves: Larvae fall into water

Development on vegetation: Larvae fall into water

Direct development: No larvae, young hatch as miniature adults

PRIMITIVE AMPHIBIAN PATTERN

*Brood pouch: Larvae or direct development

Larva metamorphose into adults

WATER OF OUT LAID EGGS EGGS LAID IN WATER

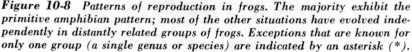

Figure 10-8 *Patterns of reproduction in frogs. The majority exhibit the primitive amphibian pattern; most of the other situations have evolved independently in distantly related groups of frogs. Exceptions that are known for only one group (a single genus or species) are indicated by an asterisk (*).*

and exploration by divergent lines of the ancestral stock, until one of them crosses the ecologic barrier into the new zone.

2. The breakthrough and shift are always rapid; otherwise they fail because of the extreme negative selection in unstable ecologic zones.

3. The new zone is always ecologically accessible, is devoid of competition, and requires a new general adaptive type for its invasion.

4. Adaptive radiation always follows the initial shift.

The origin of reptiles from amphibians demonstrates these several points.

1. Numerous divergent lines within the ancestral amphibians were taking on one or the other of primitive reptile characteristics in Pennsylvanian times.

2. The shift occurred over a relatively short period of geologic time.

3. The new zone — completely terrestrial life — was unoccupied, devoid of competition, and accessible, since the ancestral amphibians spent only part of their life on land. The principal new general adaptations that made the invasion possible were development of an impermeable skin to prevent desiccation of the adults, and the land-laid egg that allowed the young to develop on land.

4. Adaptive radiation following the shift has already been covered in detail.

THE PROCESS OF EVOLUTION   The major feature of organic evolution is divergence guided by the molding force of natural selection. Evolution at the populational level is driven by the elemental forces of mutation, selection, and drift. In the short-term view, populational evolution may be sequential, but it is always divergent in the end. Speciation, macroevolution, and megaevolution represent stages or levels in a continuum of evolution; all are driven by the elemental forces but are subject to increasingly complicated effects from other less understood forces as well. Selection becomes of greater and greater significance above the microevolutionary level.

Of greatest importance in speciation is the origin of isolation. Selection may act to produce divergence in this process whenever fragments of an originally interbreeding gene pool become spatially or reproductively isolated. New species are derived from old ones through the origin of reproductive isolation and independent microevolution.

Macroevolution and megaevolution are both extremely complex in their dimensions. They involve, in addition to the elementary forces and isolation, the following characteristics, all related to natural selection and environmental relations.

1. The taking on of new general adaptation or occupancy of a new adaptive zone.

2. The breakthrough into new zones or subzones within the new adaptive zone by development of special adaptations.

3. The loss of evolutionary flexibility, and channelization into greater and greater specialization within subzones.

4. The ecologic reinvasion of a zone or subzone when it becomes partially unoccupied because its original occupiers are now .specially adapted (ecologic replacement).

5. The irreversibility of evolution. Since each step is dependent upon the previous progressive changes, once a group is on the adaptive road, it is usually trapped in an adaptive zone and cannot reverse evolution against the direction of selection.

As an aid to thinking about the overwhelming complexity of evolution, these interactions are diagrammed (Figure 10-9) on an adaptive grid. In the words of George G. Simpson, quoted from *The Major Features of*

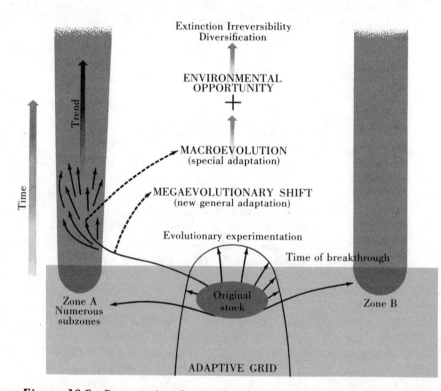

Figure 10-9 *Process of evolution above the species level. A diagrammatic representation of macroevolution and megaevolution.*

Evolution: "In adaptive radiation and in every part of the whole, wonderful history of life, all the modes and all the factors of evolution are inextricably interwoven. The total process cannot be made simple, but it can be analyzed in part. It is not understood in all its appalling intricacy, but some understanding is in our grasp, and we may trust our own powers to obtain more."

FURTHER READING

Colbert, E. H., *The Dinosaur Book*. New York: American Museum of Natural History, 1951.

Ehrlich, P. R., R. W. Holm, and P. H. Raven, *Papers on Evolution*. Boston: Little, Brown, 1969.

Goin, C. J., *Amphibian Pioneers of Terrestrial Breeding Habits*. Washington, D.C.: Annual Report of the Smithsonian Institution, 1960.

Rensch, B., *Evolution Above the Species Level*. New York: Columbia University Press, 1960.

Salthe, S. N., *Evolutionary Biology*. New York: Holt, Rinehart and Winston, 1972.

Simpson, G. G., *Horses*. New York: Oxford University Press, 1951.

Stebbins, G. L., "Adaptive radiation and trends of evolution in higher plants," *Evolutionary Biology*, vol. 1 (1967), pp. 101–142.

———, *The Basis of Progressive Evolution*. Chapel Hill: University of North Carolina Press, 1969.

———, *The Flowering Plants: Evolution Above the Species Level*. Cambridge: Belknap Press, 1974.

Tax, S. (ed.), *Evolution After Darwin: The Evolution of Life*. Chicago: University of Chicago Press, 1960.

Wake, D. B., "Comparative osteology and evolution of the lungless salamanders, Family Plethodontidae," *Memoirs of the Southern California Academy of Sciences*, vol. 4 (1966), pp. 1–111.

chapter 11

On the Origin and Evolution of Life

Since time immemorial human beings have pondered their origins and have attempted to explain the creation of the cosmos, life, the social order, and themselves as sentient creatures. One of the major triumphs of modern science is that it has revealed to us that the energy-releasing and emotion-invoking myths which form the sources of all culturally unifying religions are reflections or projections of inner psychic processes onto the external world of nature. Science has demonstrated that the world of nature with its prodigous display of phenomenon, which we can never fully hope to comprehend, can be understood to a substantial extent through our senses and their instruments and can be translated to our mind.

Man's search for certainty has often led

him to regard myth as an accurate and factual description of the cosmos, life, and man. Literal interpretations of myths which corrupt the myth provide security by establishing an unchanging and unchallengeable explanation of events and in the extreme case substitute dogma for reality in the name of absolute truth. The role of science, on the other hand, has been to create a new vision of the world, based on observation, experiment, hypothesis formation and testing, combined with constant revision of previously accepted ideas. Science does not and cannot pretend to be true or final in any absolute sense. It is an organization of tentative truths that are today's scientific facts, laws, and theories, but which will surely be modified tomorrow into new and different explanations by further research. Science thus provides a world of change, new ideas, new things, new and previously unrealized horizons, and continuing transformations, just like life and evolution themselves.

Because of its tremendous power of explanation of the external world, science and many scientists may often seem to ignore the inner world that each human being senses within himself, where his most creative aspects, feelings, imagery, poetry, and myth constantly arise. Fear and distrust of science is often the result, especially when some scientists in that all too human search for certainty become confused about scientific knowledge and speak of universal and unchanging truths.

It is understandable, then, that in a time of such continuous, radical, and boundless search for the ever-changing truth, certain religious people feel threatened by the concepts of life origins and evolution that are accepted by most scientists. The certainty and the security provided such persons by a literal acceptance of the description of the creation of the earth and life established in biblical accounts as historical fact, and adherence to the injunctions of their formal religion, must indeed seem generally preferable to facing the constant changes in scientific thought and modern life. In recent years many such individuals, mostly from fundamentalist, protestant Christian sects and including a number of scientists, have actively attempted to have a "creationist" hypothesis of life origins based on biblical accounts in the book of Genesis, accepted as an alternate to the current scientific thesis of evolutionary change. As we shall see, the creationist view cannot be tested and is therefore not scientifically viable. Most previous attempts by religious groups to refute evolutionary theory have centered on the processes of speciation. In contrast the creationists concentrate on ideas relating to the origin of life, the most significant and least understood aspect of the evolutionary process.

At another level, the creationists are saying something else. They sense the tendency of certain scientific fashions and scientists to take dogmatic positions, to believe that everything in the universe may be explained by a few principles, to ignore phenomena that cannot be tested by these principles, and after a while to pretend the phenomena no longer exist. Thus the extreme reductionist position in scientific materialism claims that everything can be explained in terms of physical laws that involve matter and energy. Phenomena that cannot be explained in these terms then become unworthy of scientific study, and within a short time the phenomena do not exist. In this regard scientific reductionism has merely changed the name of the principle of existence from God, the supreme reality, the universal mind, or the ultimate ground of being, and called it matter or energy, which explains nothing. The implied challenge to science is to allow doubt and uncertainty fair play while pursuing new forms and truth, even though old theories still hold sway.

POSSIBILITIES FOR THE ORIGIN OF LIFE ON EARTH Four principal hypotheses regarding the origin of life on planet III of our solar system are currently advanced in Western theology, philosophy, and science. Other hypotheses are usually variants of the basic four or are strictly mythologic. The ancient belief that many living creatures arise spontaneously from nonliving material (bats from mud, flies from meat, and so forth) has been demonstrated repeatedly not to be true. As far as is known, all living creatures currently on earth arose from other organisms of the same kind. Some scientists have speculated that life was originally brought to earth accidentally or purposely from other planets in our galaxy. Although the evidence seems conclusive that life on earth originated on earth, if the view of an extraterrestrial origin is accepted, some sort of hypothesis to explain the ultimate origin of life would still be required. The idea that some one divinity or several gods are constantly placing new kinds of organisms on earth, although a feature of some myths of creation, seems to have no scientific validity. Of the great number of creation myths believed by various peoples, only the Judeo-Christian biblical account is used by the creationists as an alternative to evolutionary explanations.

1. The origin of life is a unique event directed by divine forces incomprehensible to the human mind. Many proponents of this view accept this position based on the revelatory nature of the Old Testament; others marshall evidence demonstrating possible or imagined flaws in the alternate hypotheses to conclude that this concept is the only viable one. Scientifically, this hypothesis cannot be tested, since no evidence in the usual sense will cause its serious adherents to question divine authority.

2. Life originated as a highly unlikely event under a unique set of environmental circumstances during the geologic evolution of planet III. Discoveries of primitive fossil prokaryotes in rocks about 3.1 billion years

old and current knowledge of earth geophysics and geochemistry make this hypothesis very attractive.

3. Life is but one example of a general physical principle of the organization of matter under the flow of energy. Consequently studies on the physicochemistry of many different far-from-equilibrium systems could lead to duplication of life in the laboratory. This view provides a scientific basis for experimental studies related to the processes that may have given rise to life in the past.

4. Life originated under the control of an as-yet-undiscovered scientific principle, and the operation of this principle together with evolution directed the formulation of living things.

Although some scientists favor the latter hypothesis as the most likely of the four, its study and testing has as yet not been undertaken by serious students of the problem. Most studies are centered on physicochemical shifts from nonliving to living systems, which test various aspects of hypotheses 2 and 3.

POSSIBLE MODEL FOR THE ORIGIN OF LIFE
In this section I will describe the most acceptable current model for the origin of living systems. It is a model based on the best available knowledge of earth history, geophysics, geochemistry, and molecular activities and transformations. The model includes elements from the studies basic to hypothesis 2 and from the experiments supporting hypothesis 3. It is a speculative description similar to accounts by several other authors. Almost every point in the discussion can be argued from several points of view, and scientific testing of many of the inherent assumptions may never be possible. Scientists, as you must understand by now, make intellectual progress by proposing conceptual structures that go far beyond the available data. Such an explanation usually provides a basis for testing ideas, new data, and formulations in a more rigorous and/or experimental fashion so that the hypothesis may undergo continuous growth and change. Unfortunately, since scientists are human too, what were originally useful conjectures, sometimes through repetition begin to be accepted as fact (proof by repetition). The model presented here is almost of that kind — since its essential framework has been discussed in many publications and beginning textbooks. In some cases the hypothesis is presented in such a way as to imply that scientists know what happened when life was originating. The processes, sequences of events, and their timing are then described as though they actually occurred, instead of being an imaginative reconstruction of what

might have occurred. In reality, the model is a very rough approximation of what might have transpired, with the gaps filled in by logical and plausible interpolations. As such, it is informative, perhaps thought-provoking, and certainly fun to play with, as long as no one takes it too seriously. It is very little in terms of hard data and rigorous testing, but it is all we have (Figure 11-1).

As previously outlined (Chapter 1), our solar system arose about 6 billions of years ago through the cooling and condensation of a huge gaseous cloud. Subsequent cooling led to the gradual evolution of the planet earth by 4 to 4.5 billions of years ago into an environment where

---

Figure 11-1 Stages in the evolution of life on earth. (A) *Stage of the organogenic molecules: formation of simple organic molecules through the spontaneous combination of gases from the then reducing atmosphere of hydrogen (H_2), ammonia (NH_3), methane (CH_4), and water vapor (H_2O) through the input of electrical and ultraviolet light energy in freshwater. Time: about 4.5 billion years ago.* (B) *Stage of the protobionts: formation of complex aggregates of a variety of organic molecules (coacervates) or liner organic molecular complexes (polymers); the earliest kind of life may have been formed by a symbiotic association of a membrane-enclosed coacervate and a polymer as shown (1). Time: about 4 to 3.5 billion years ago.* (C) *Stage of early life: formation and diversification of the first living systems as prokaryotic primary heterotrophs dependent on anaerobic cellular respiration to release energy stored in organic molecules; the waste product of this process is carbon dioxide (CO_2) that began to build up in the earth's water and atmosphere. Time: between 3.5 and 3.4 billions of years ago.* (D) *Stage of first plants: appearance and diversification of life forms (autotrophs) capable of capturing and storing the sun's energy through a special green catalyst, chlorophyll, in the process of photosynthesis; this process uses carbon dioxide (CO_2) and water (H_2O) to manufacture glucose with free oxygen (O_2) given off as a waste production; the atmosphere now becomes an oxidizing one similar to the atmosphere of today; free oxygen molecules (O_2) react with one another in the upper atmosphere to form an ozone (O_3) shield that screens out ultraviolet rays; the deleterious effect of these rays on living material had meant that up to now life survived by staying 5 to 10 meters below the water's surface; with the increase in available oxygen, life invented a new and more efficient system of cellular respiration (aerobic) in which oxygen combines with glucose in the cell to release energy and gives off carbon dioxide and water as by-products. Time: About 3.2 billion years ago.* (E) *Stage of first animals: appearance and diversification of secondary heterotrophs that feed on other organisms or organismal products and use the efficient means of aerobic respiration to release energy for an active, free-living motile life. Time: about 3 billion years ago.* (F) *Stage of modernization: appearance and diversification of multicellular plants and animals, invasion of land and general trend to increasing complexity of life. Throughout each of the later stages (C–F) the metabolic pathway of anaerobic respiration was continued by a few organisms, but once photosynthesis and aerobic respiration pathways appeared they predominated and led to diversification and complexity in a continuous radiating process during the last 2 billion years.*

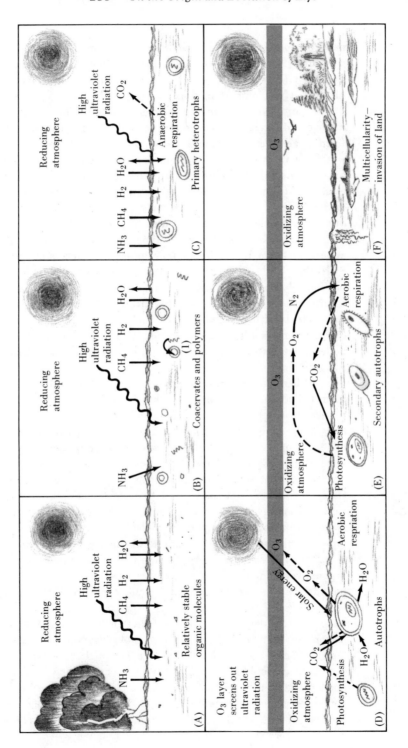

(A) Reducing atmosphere — High ultraviolet radiation — NH₃ CH₄ H₂ H₂O — Relatively stable organic molecules

(B) Reducing atmosphere — High ultraviolet radiation — NH₃ CH₄ H₂ H₂O — (I) — Coacervates and polymers

(C) Reducing atmosphere — High ultraviolet radiation — NH₃ CH₄ H₂ H₂O — CO₂ — Anaerobic respiration — Primary heterotrophs

(D) O₃ layer screens out ultraviolet radiation — Oxidizing atmosphere — O₃ — Solar energy — O₂ — Aerobic respriation — Photosynthesis — CO₂ — H₂O — H₂O — Autotrophs

(E) Oxidizing atmosphere — O₃ — N₂ — O₂ — Aerobic respiration — CO₂ — Photosynthesis — Secondary autotrophs

(F) Oxidizing atmosphere — O₃ — Multicellularity— invasion of land

new and complex molecules could be formed in the extensive areas covered by water. The atmosphere of the earth at that time contained much more hydrogen (H_2) than it does now, and no or very little free oxygen (O_2), nitrogen (N_2), or carbon dioxide (CO_2), the gases that form by volume 21, 78, and 0.03 percent, respectively, of the atmosphere today. Instead nitrogen, oxygen, and carbon were joined with hydrogen molecules to form a gaseous atmosphere of ammonia (NH_3), methane (CH_4), and water (H_2O) vapor. Surface temperatures were generally higher than at present, and solar radiation, especially the ultraviolet portion, and perhaps electrical storms and thermal activity in the earth's crust provided sources of free energy that may have contributed to the formation of complex organic molecules during these times (Figure 11-1).

To test this possibility, in 1953 Stanley L. Miller built an airtight apparatus through which the four gases, methane, ammonia, water, and hydrogen, could be circulated past electrodes that made it possible to pass an electrical charge into the gas mixture. The gases were allowed to circulate for a week, energized by the electrical sparks. At that time the contents of the chamber were analyzed and an astounding number of organic compounds were discovered. Among the spontaneously produced molecules were several amino acids, the components of protein, the *sine qua non* molecules of life.

Miller also ran two controls. (1) He sterilized the gas mixture at high temperatures for eighteen hours before turning on the electrical power source; the yield of organic molecules was similar in this case to that in the original experiment. (2) He ran an experiment just like the original but without providing electrical energy; there was no significant yield of new organic molecules.

Since that time Miller's experiment has been extensively duplicated. Other researchers have subjected hydrogen and methane atmospheres to electrical energy to produce a long list of organic compounds, including most significantly the important molecules that make up the complementary organic bases in DNA and RNA molecules. Still other researchers using other gases and energy sources have synthesized many additional organic compounds. The conclusion is inescapable: Complex organic compounds may be synthesized from simpler molecules which are subjected to a source of energy.

Thus it seems almost certain that in the period between 3.5 and 4 billions of years ago, large organic molecules were being synthesized in the freshwater oceans of the earth. It is thought that any large molecules so produced would survive for very long periods and have a strong likelihood of interactions with other such molecules. Of course these kinds of molecules could not survive for long today, even if they were produced, because of the activities of microorganisms that would break down the molecules very shortly after they appeared.

It now seems highly probable that under these conditions and over a very long time the organic compounds in the shallows of great seas combined and recombined with others to produce over a billion years more and more complex and stable aggregates of large molecules. It has been suggested by many researchers that among those which inevitably arose were complex colloid aggregates (coacervates) some of which later came to be bounded by an organic membrane. Such membranes have the ability to select the kinds of molecules that will be added to the aggregate. Internally, the aggregates or protobionts (precursors to life) are thought to have become increasingly ordered and through this orderliness to regulate the type of chemical reactions that could occur, as well as providing sites for their occurrence. From a probabilistic point of view there were doubtless many different kinds of these aggregates, but laboratory studies of such systems show that those with the more efficient and coordinated reaction components tend to grow and maintain themselves for longer periods than coacervate systems with disorganized constituents. As protobionts assumed greater chemical and structural complexity, those with the ability to obtain a steady supply of energy and chemical building blocks from the surrounding sea and later those developing chemical systems that manufactured some of these materials internally would have tremendous advantages over the others.

Other scientists believe that during this time the various organic compounds (that is, amino acids, sugars, nucleic acid bases) became organized into longer and more complex chains or polymers (that is, proteins, polysaccharides, and nucleic acids) and were not associated with coacervates. In the benign earth sea with an abundance of nutrients and few consumer molecules, one can imagine a great variety of polymers developing. Somehow, somewhere certain of these long molecules organized into advantageous combinations that had the capacity to produce complementary copies of themselves. In order for these systems to have significance at some point in time, the polymers would have to develop the ability to bring about the formation of organic catalysts. Because of these requirements one is almost forced to believe that the successful polymers were DNA-like or RNA-like molecules which gradually acquired self-duplicating and information-transferring features that allowed a limited capacity to organize amino acids into polypeptide chains with catalytic properties. Later, by some unexplained means these information-carrying molecules wrapped themselves in a membrane envelope to become the first living thing.

A third view combines the two above and suggests that DNA or RNA molecules became associated symbiotically with a membrane-enclosed coacervate to become an advanced protobiont. Such an organism comes close to the basic definition of life: The colloid organic mixture of the coacervate provides a milieu for autosynthesis under control of the DNA or RNA molecules. In turn these information-storing and -transferring molecules provide the basis for reproduction and evolutionary change.

Although the development of protobionts as entities containing chemical reaction sites is an important step toward life, each protobiont stood as a unique experiment in chemical evolution without the means for increasing the frequency of its unique features among the diverse systems in the ancient sea. The next series of steps in evolution are then the important ones, because they lead to systems having the key feature of life: autosynthesis (self-regulated conversion and production of energy), autocatalysis (self-duplicating reproduction), and evolution (mutability). Life as opposed to prelife is characterized by the coupling of these three processes. It seems obvious that in living forms all three are intimately related to the characteristics and activities of the nucleic acids (DNA and RNA) which regulate proteinaceous enzyme synthesis and are self-duplicating and mutable transferers of genetic information. Prelife forms, even those in the transitional phases discussed here, lack at least one of these features. A major question with no easy answer relating to the origin of living entities from protobionts is, How did the latter acquire the DNA or RNA structure that is the essence of life? Perhaps the nucleic acids evolved as one of the constituents within the coacervate aggregates, first as rather short and simple chains of molecules that in time become longer and more complex. When this process was completed (no one knows how, when, or how long it might have taken), the first living organism came into being.

This development involved the following events, but we have no basis for knowing even a probable sequence.

1. Acquisition of autosynthetic properties (metabolically "living")
   a. Thermodynamic component — during chemical evolution one to several kinds of molecules that can be broken down to release regulated amounts of free energy must have been incorporated into the protobiont
   b. Kinetic component — origin of catalysts that mobilize the energy contained in the storage molecules and regulate the rate of reactions in the system; these are mostly *very* specific enzyme proteins in extant organisms that *cannot* moderate a variety of reactions and are composed of unique sequences of amino acids determined by the nucleic acids; recent evidence suggests that the original catalysts were inorganic compounds
2. Acquisition of autocatalytic and mutable properties (genetically "living")
   a. Information storage — encoding of exact template data for synthesis of specific amino acid sequence for specific proteins

    b. Mutation potential—ability for spontaneous appearance of novel sequences to provide new varieties of proteins

    c. Information transfer — transcription and translation (Figure 1-7) of information into protein

    d. Precise replication — biological reproduction (Figure 2-5) of the stored information to be passed on to the next generation

Of these latter features only a and b are functions of the nucleic acids themselves. The other two processes (c and d) require participation of specific energy sources (thermodynamic component) and enzymes (catalytic component). It is this coupling of nucleic acid activity with protein enzymatic regulation that creates the greatest experimental and conceptual gap in any explanation of the appearance of life. Neither system (protein nor nucleic acid) alone meets the requisites as a unit to which the other could be added later to form life. Both are extremely complex and at the same time so closely integrated in action and detail that it is difficult to conceive of either evolving in the absence of the other. This kind of difficulty leaves the question of the process of life origins still a mystery and challenges biologists to think anew on alternate models and their implications. Contrary to the hopes of many scientists, it seems we are a very long way from understanding the most basic facets of the origin and nature of life and even further from the day life may be synthesized by human beings in the laboratory.

When the first living system appeared, not only the earth but also the cosmos was changed. Since the nucleic acids have the capacity for self-replication and novel mutability, life has both the ability to perpetuate its own kind and, under the guidance of natural selection, to ensure the increase in its unique features among the populations of the world. From the appearance of the earliest living system onward, survival and expansion of life were concerned with the coming generation. The nucleic acids make it possible for life to transmit a life plan to the next generation. Mutation provides variety in available life plans. Natural selection acts to encourage the survival and expansion of the best available life plan, and in this way the nucleic acids provide life with the means to learn from the experience of existence and to pass this knowledge to its descendants. The ultimate significance of the origin of life is the appearance in the cosmos of a system that by inductive processes passes on to its species as encoded in its DNA or RNA all that its lineage has learned. Life transformed evolution so that what was now important was not the future of a single entity, but survival of its offspring. The information-storing and self-replicating functions of DNA and RNA assure the unique feature of organic evolution: survival through preservation of the wisdom of the past, amid revolution constantly encouraging change.

EARLY EVOLUTION OF LIFE The earliest forms of life must have been very small microorganisms that lived humbly and inconspicuously in the warm waters of the earth. Perhaps at first hardly distinguishable from some of the protobionts but so much more efficient that, within a short time, nonliving organic aggregates were all but eliminated by the foraging of living creatures. The earliest kinds of life seem, like most of the living bacteria, to have been *heterotrophic* in their metabolism (that is, they must take in energy sources from the environment). These early creatures apparently broke down various kinds of organic molecules, took them into their cells, and converted them into glucose. Then by *anaerobic* (not requiring oxygen) *respiration*, energy in the chemical bonds of the glucose is released to provide energy for the cell in the form of a thermodynamic molecule, adenosine triphosphate (ATP) with waste products of ethyl alcohol and carbon dioxide released to the environment as follows.

$$C_6H_{12}O_6 \rightarrow 2CH_3CH_2OH + CO_2\uparrow + 2\ ATP$$

glucose   ethyl alcohol   high-energy
                                     source

Over a long period this process led to the addition of substantial amounts of free CO_2 gas to the water and atmosphere of the ancient world (Figure 11-1).

Much later, probably about a million years after life's appearance, a new kind of life emerged. In the cells of this new kind of life, because of a unique evolutionary invention, a green substance called *chlorophyll*, it was able to fix the sun's energy into glucose molecules, by a process called *photosynthesis*. These kinds of organisms are called *autotrophs*, because they fix their own energy and are not dependent upon finding organic molecules in their environment as energy sources. The process of photosynthesis utilizes solar energy, water, and carbon dioxide in the presence of chlorophyll as a catalyst, to manufacture glucose as follows.

$$6CO_2 + 6H_2O + chlorophyll \xrightarrow{\text{light energy}} C_6H_{12}O_6 + 6O_2\uparrow$$

Free gaseous oxygen is given off as a waste product in this process. For these first green cells carbon dioxide became a utilizable resource. The free oxygen given off during this efficient process of energy fixation changed the atmosphere of earth. Hydrogen and methane are turned into water and carbon dioxide by free oxygen. In addition, free oxygen in water breaks down organic molecules to eliminate many kinds of possible coacervates and life forms that prey directly on them. Furthermore, free oxygen became available for the first time for a new form of effective energy release that is many times more effective than fermentation.

The process, *aerobic respiration* (respiration using oxygen), unlike fermentation, which yields only a minimal amount of energy from glucose, releases in a regulated stepwise process almost 100 percent of a glucose molecule's energy. The process is summarized below.

$$C_6H_{12}O_6 + 6O_2 \longrightarrow 6CO_2 + 6H_2O + 36 \text{ ATP}$$

In the reaction carbon dioxide and water are given off as waste products, and 36 molecules of the thermodynamic molecule (ATP) with the stored energy are formed. The energy in the ATPs can be used to carry on work anywhere in the organism.

Almost all living organisms, except most bacteria, utilize aerobic respiration. In green plants energy is produced by photosynthesis and energy is released by aerobic respiration. Most of the myriad heterotrophs (principally animals) now present on earth are regarded as descendants of autotrophs that lost the photosynthetic capacity secondarily and while maintaining aerobic respiration became active hunters and gatherers of organic molecules, usually concentrated in the form of living cells and bodies of other organisms (Figure 11-1). In terms of major metabolic pathways the story of life is essentially complete once earth became dominated by autotrophs and secondary heterotrophs in the form of a whole series of unicellular plants and animals by at least 3 billion years ago (Figure 11-2).

Other significant events in the history of life are those discussed below.

1. Establishment of cooperation rather than competition as life's crucial technique for survival; the interdependence of life and its mutually benefiting effects on environment and other organisms is demonstrated by the role of primary heterotrophs as essential to the evolution of autotrophs and of the latter as energy sources for secondary heterotrophs (Figure 11-1)

2. Appearance of sexual reproduction that biotically requires the cooperation of two parents and provides the offspring with more tools (variation) in their struggle for survival by creating novel recombination of genetic materials

3. Taking on of multicellularity which requires elaborate and intimate cooperative self-organizing and self-regulating by a functioning individual made up of several to many million cells; a process that allows for greater size, activity, complexity, and coordinated division of labor among the cells; complex multicellular organisms probably appeared about one billion years ago

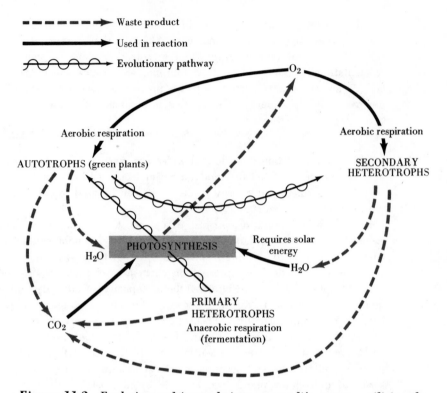

Figure 11-2 *Evolution and interrelations among life systems utilizing the three important metabolic pathways of anaerobic cellular respiration, photosynthesis, and aerobic cellular respiration. The earliest living systems are thought to have been primary (anaerobic) heterotrophs.*

4. Association of individual death with all higher forms of life; unicellular organisms may go on dividing indefinitely but sexual reproduction and/or multicellularity has brought life to a point where the sex cells containing the nucleic acid plans for the next generation are specialized to assure genetic immortality, but it is now the usual event for each individual to die as part of the life cycle

5. Multiple and repeated invasion of the land, where the rich oxygen source of air encourages higher levels of energy use with concomitant increased activity and complexity, primarily by multicellular plants and animals

6. Gradual evolution in multicellular animals of the origin of feelings (awareness) and motivation

Important long-term evolutionary trends exemplified by the 3.5 billion years of life seem to emphasize the following.

1. Increasing intracellular and intercellular, intrapopulational and interpopulational cooperation as a prerequisite for the spectrum of living forms and relationships

2. Increasing complexity, adaptability, and activity

3. Increasing flexibility and loosening of the rigid determination of the inorganic world and heredity so that motivation, choice, value as well as their outcomes, freedom and responsibility, become possible.

4. Increasing learning and heightening levels of consciousness

And then in man the limits of biology are transcended.

FURTHER READING

Campbell, J., *Myths to Live By*. New York: Viking, 1972.

Handler, P. (ed.), *Biology and the Future of Man*. New York: Oxford University Press, 1970.

Long, C. H., *Alpha. The Myths of Creation*. New York: Collier, 1969.

Miller, S. L., *The Origins of Life on Earth*. New York: Prentice-Hall, 1974.

chapter 12

The Rise of Man

Among the most controversial aspects of Charles Darwin's ideas on evolution was his recognition that man was closely allied to other animals and appeared to have evolved from them. Although detailed fossil evidence was lacking, Darwin freely predicted on the basis of morphology and behavior that stages intermediate between apes and men would ultimately be discovered. During the century since the publication of *The Origin of Species,* this concept has been repeatedly attacked on nonscientific and metaphysical grounds. Concurrently, however, fossil materials of apes, apemen, and men have been gathered from a wide variety of sources, and the cumulative evidence, particularly recent discoveries, unequivocally supports the theory of human origin from the higher apes.

Human evolution as seen in its basic outlines does not involve processes or mechanisms unique in the biotic world. Indeed, the fact that our historical development exhibits the same general patterns of linear and divergent evolution characteristic of all life is compelling evidence of man's descent from other organisms. Man is, of course, a unique product of evolutionary forces and has attributes not found in other species; but so is every other species of living organism unique in its particular characteristics and evolutionary development. Human beings are interested in human evolution not because of any special evolutionary forces responsible for our origin and development, but because man's egocentric anthropomorphism is one of our biological attributes. But, of course, man is the only creature capable of such interest. In some individuals egocentrism is carried to the extreme in a complete denial of our biological relations; the statement is heard that we are so peculiar that we cannot be the products of biological development and cannot be descended from other animals. Unfortunately, such statements are not based on evaluation of the evidence but rather on emotion or metaphysics. As will be readily apparent in the discussion below, the crucial evidence of human evolution and the essential outlines of man's evolutionary progress are overwhelmingly convincing to all men with open minds. Those who continue to insist that the recognition of our evolution from other animals somehow debases us, or in some unknown manner destroys the biological attributes that have made us successful and unique evolutionary creations, prefer to disregard the overwhelming validity of the evidence that is now in existence.

THE EVOLUTIONARY HISTORY OF MAN   It has long been recognized that man as a species is related to a rather diverse group of mammals, placed by modern biologists in the order Primates. This order is regarded as being among the more primitive groups of placental mammals and is characterized by retention of many generalized features that have taken on extreme specializations in more highly evolved mammal orders. The most significant structural features of the primates, and those to be expected in any proposed stock ancestral to man, are the following.

1. Basically arboreal habits, some forms becoming terrestrial
2. Limbs, hands, and feet adapted for arboreal existence, with opposable thumbs and big toes as modifications for grasping branches
3. Vision and hearing the dominant special senses, with enlargement of appropriate areas of the brain for sensory reception from eyes and ears

4. Concurrent enlargement of the braincase to make room for expanded cerebral portions of the brain

5. Generalized dentition and food habits

Living primates may be placed into two major groups and nine subgroups or families (Figure 12-1).

I. Prosimii: primitive primates
    Lemuridae: lemurs — Madagascar
    Daubentoniidae: aye-ayes — Madagascar
    Lorisidae: lorises and galagos — tropical Africa and Asia
    Tarsiidae: tarsiers — East Indies and Philippines
II. Anthropoidea: advanced primates
    Cebidae: New World monkeys — tropical America
    Callithricidae: marmosets — tropical America
    Cercopithecidae: Old World monkeys — Africa and Asia
    Pongidae: apes — tropical Africa and Asia
    Hominidae: men — cosmopolitan

The course of evolution among the primates other than man may be sketched in broad outline on the basis of the structure of living forms and a fragmentary but convincing fossil record. The very earliest primates are closely allied with the generalized basal stock of placental mammals, a group called the *insectivores*. The insectivores are usually small, active creatures and include, among living forms, the shrews, moles, and hedgehogs. The most primitive known primates were apparently similar to shrews in structure and resembled the living tree shrews. Fossil remains of these forms are known from the Paleocene, indicating the presence of possible primate ancestors about 70 million years ago.

More highly evolved members of the order, including fossil lemurs, tarsiers, and monkeys, have been recovered from Eocene deposits in both Eurasia and America. The first two groups appear to have originated from shrewlike ancestors. Monkeys, in turn, seem to have been derived from rather generalized tarsiers, which lacked the specializations found in modern forms. Fossil materials from the Oligocene (40 million years ago) form convincing connecting links between the Old World monkeys and the apes. Apparently the most primitive apes were small, monkey-sized animals. Numerous ape remains are known from Miocene and Pliocene times. Significantly, although they are definitely apelike in most regards, all these ancient apes exhibit few of the extreme specialization of the modern gibbon, orangutan, chimpanzee, or gorilla, but have characteristics somewhat similar to those found in man. Certain of these Mio-Pliocene apes (*Dryopithecus*) appear to have inhabited forests and, like the chimpanzee and gorilla, to have spent some of their time on the ground and some in tree climbing. These forms are regarded as the probable ancestors of the

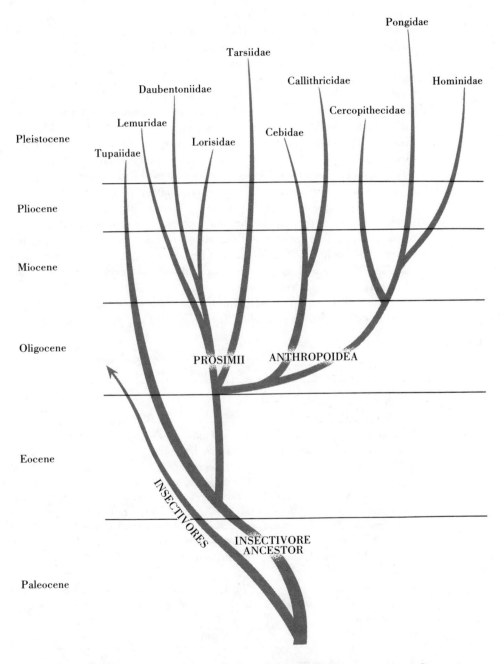

Figure 12-1 *History of the primates. Only the names of living families are included here. A number of now extinct families are omitted from the diagram.*

modern great apes. A second group (*Ramapithecus*), now known from eastern Africa, India, and Hungary, lived in open country and is presumed to have walked on all fours, although evidence is based solely on skull fragments. In most features of the skull and dentition, the latter animal resembles man. Modern great apes are more specialized than man in many respects, particularly in those features related to arboreal habits, and the fossil forms share with man the more generalized primitive conditions. Even though living apes are less humanlike in many ways than are their extinct primitive ancestors, it has been obvious to the majority of biologists for 200 years that men and apes are closely allied. The rather recent discoveries of primitive fossil apes indicate an even closer affinity than was originally suspected. The similarities between man and the manlike or anthropoid apes have led to a general theory of a common ancestry of the two groups, a theory enhanced by the striking similarities between primitive fossil apes and the human species. The crucial fossil evidence of man's origin, however, has been uncovered in a series of exciting discoveries beginning in South Africa fifty years ago and accelerating with finds in Ethiopia and in central and South Africa during the past decade (Table 12-1).

The most important structural differences between apes and man are summarized below as a basis for evaluating the position of the newly

Table 12-1   Proto-men, Near-men, and Men
(Family Hominidae)

Name	Area	Age in Years Ago
1. *Ramapithecus* (*Kenya-pithecus*)—2 species	India, Pakistan, East Africa, Hungary	14–9 million
2. *Australopithecus africanus*	East and South Africa, Ethiopia, China, Java	5.5 million–700,000
(*Australopithecus robustus,** Homo africanus, Homo habilis, Homo modjo-kertensis, Homo transvaalensis,** Meganthropus africanus, Meganthropus palaeojavanicus, Paranthropus crassidens, Paranthropus robustus,** Plesian-thropus transvaalensis, Telanthropus capensis*)		
3. *Australopithecus boisei* (*Zinjanthropus boisei*)	East Africa, Ethiopia	3–1.2 million
4. *Homo erectus* Includes *Homo habilis** of most authors	North and East Africa, Europe, China, Java	2.3 million–300,000
5. *Homo sapiens*	Worldwide	300,000–0

Names in parentheses are important synonyms.
An asterisk (*) indicates a name often used in other books for this species.

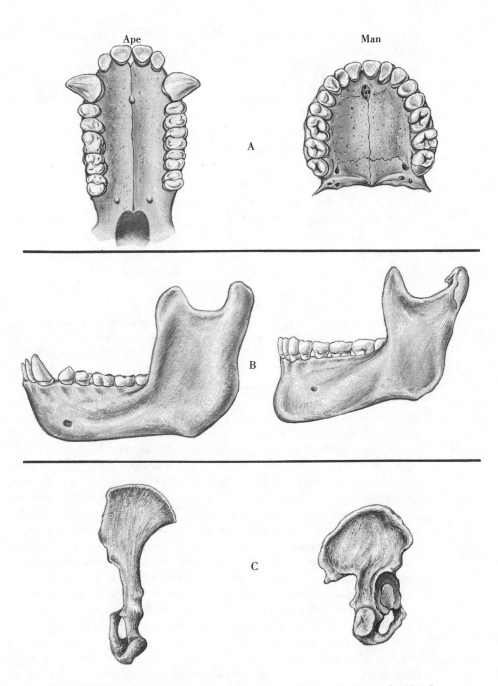

Figure 12-2 *Comparison of apes and men:* (A) *roof of mouth;* (B) *lower jaw;* (C) *pelvis (lateral view).*

discovered fossil primates so critical to our understanding of human origins (Figure 12-2).

<table>
<tr><td align="center">APES</td><td align="center">MAN</td></tr>
<tr><td>1. Cranium expanded; maximum brain size 750 cc</td><td>1. Cranium greatly expanded; maximum brain size 2200 cm</td></tr>
<tr><td>2. Occipital condyles posterior</td><td>2. Occipital condyles anterior</td></tr>
<tr><td>3. Strong nuchal crest, reaching high up onto back of skull</td><td>3. Low nuchal crest, not extending high up onto back of skull</td></tr>
<tr><td>4. Palate long</td><td>4. Palate reduced</td></tr>
<tr><td>5. Incisor and canine teeth large</td><td>5. Incisor and canine teeth reduced</td></tr>
<tr><td>6. Anterior premolar in lower jaw strong and pointed</td><td>6. Anterior premolar in lower jaw small and bicuspid</td></tr>
<tr><td>7. Pelvis narrow and elongated</td><td>7. Pelvis broad and flattened</td></tr>
<tr><td>8. Limited use and no manufacture of tools</td><td>8. Extensive use and manufacture of tools</td></tr>
</table>

Correlated with man's expanded cranium and large brain size are his ability to reason, his fine memory, his range and intensity of awareness and emotional repertoire, his learning and flexibility of response, and his use of language, all of limited significance among the apes. The features of condyle location, nuchal crest development, and pelvic structure are associated with bipedal locomotion. The first two are related to the position of the head and its muscular supports in an upright stance; the third with support of the body and muscular attachments for the hind limbs in bipedal locomotion. Apes occasionally are bipedal, but normally walk on all fours when on the ground. The structural differences between ape and men in palatal and dental characteristics appear to correlate with food habits. Man is more thoroughly omnivorous than the apes and is unable to kill animal prey with his inadequate teeth and jaws. Tool manipulation and construction in man are made possible by an upright stance that frees the forelimbs for uses other than locomotion and by the large brain centers devoted to manual control. It seems likely that this posture is also responsible, together with stereoscopic (three-dimensional) color vision, which man shares with other higher primates (marmosets, monkeys, and apes), for enlarging man's vistas and encouraging the new perceptions of his environment that contributed substantially to increasing awareness. Equally important, the upright stance frees the mouth for sound making. Essentially, the elevated head, positioned far above the ground, became a center of sensory reception, perceptional activity, and social communication — to open new horizons and evolutionary possibilities.

These broad correlations between brain size, bipedalism, tool manipulation, increasing consciousness, communication, and associated features suggest that the characteristics evolved more or less as a unit, with brain size increase following the development of the others. Strong positive selection for all the characteristics essential to man's dominance of his environment probably explains the rapid shift from ape to man during the preceding 5 million years.

Although the similarities in many features and the obvious relationship between apes and modern man supported the theory of common ancestry, the differences between the two groups in the essentials discussed above had still to be explained. Among biologists in the early part of the present century, these differences provided some basis for doubting the evolution of man from an apelike stock. Then a series of extremely important finds gradually filled the gap between the two groups. The first of the critical discoveries of new fossil material was made in 1925 by Professor Raymond Dart in Bechuanaland, South Africa. Dart's animal, which he called *Australopithecus* (southern ape), consisted of a single skull that combined an unexpected number of human and ape characteristics. In succeeding years additional skulls, jaws, and other essential anatomical parts of Dart's animal were discovered in other areas of South Africa. These forms are similar to primitive apes in certain characteristics and to man in many others. The most ancient of these near-men lived in the Pliocene, about 5 million years ago, and some of them persisted until as recently as 700,000 years ago. The African near-men share the following significant structural features (Figure 12-3).

1. Cranium expanded; brain size about 500 cc
2. Occipital condyles anterior
3. Nuchal crest reduced
4. Palate reduced, but longer than in man
5. Incisor and canine teeth reduced
6. Anterior premolar in lower jaw small and bicuspid
7. Pelvis broad and flattened

On the basis of these characteristics the near-men of southern Africa resemble apes in cranial size and man in the structure of the condyles, nuchal crest, pelvis, and dentition. In the nature of the palate the near-men are intermediate between apes and man, but clearly tend toward the human condition. In addition, some of these primates used pebbles of various sorts for killing their animal prey, and others appear to have manufactured and used crude stonecutters and/or scrapers.

Interpretation of the preserved remains of the southern near-men allows us to compare their general features with those of modern man (Figure 12-4). The relatively small cranium and brain capacity suggest an order of intelligence lower than that of man but higher than that of apes. The characteristics of condyle location, nuchal chest development,

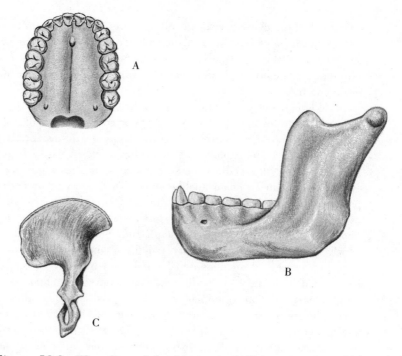

Figure 12-3   *The salient skeletal features of African near-men:* (A) *roof of mouth;* (B) *lower jaw;* (C) *pelvis (lateral view).*

and pelvis are conclusive evidences of bipedal locomotion and upright posture. The palatal and dentition conditions in the near-men suggest that, like other men, these species were incapable of killing large animal prey without tools. Although apelike in many ways, these small-brained, bipedal African forms are obviously allied more closely to man than to any other animal and form a link between him and the apes. In certain respects these men resemble more closely the primitive fossil apes previously discussed than any of the living forms. Biologists now agree generally that the South African near-men represent an early stage in human evolutionary differentiation and may be referred to the family Hominidae.

Recent discoveries in central Africa and reinterpretation of available material of near-men from South Africa reinforce the concept of human descent from these ancestors. Apparently by late Pliocene time (3 million years ago) two species of near-men had become established. One species, *Australopithecus boisei*, the larger of the two, was an essentially herbivorous form that seems to have failed to develop a strong use of tools. The second form, *Australopithecus africanus*, apparently was omnivorous. This species at least by 2 million years ago became capable of making crude stone tools with jagged cutting edges that were probably

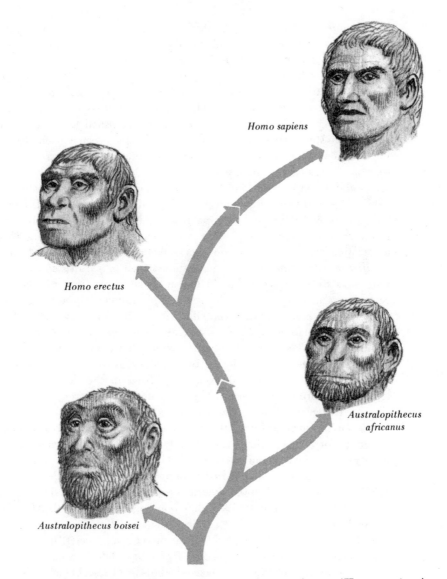

Homo sapiens

Homo erectus

*Australopithecus
africanus*

Australopithecus boisei

Figure 12-4 *Significant stages in the evolution of man* (Homo sapiens).

used to kill and skin small animals. The two forms existed together for several hundreds of thousands of years until about 1.2 million years ago. Subsequent human evolution is based on the characteristics of *A. africanus*. The features of this form indicate that tool-making abilities appeared in human evolution before the tremendous enlargement of the brain and cranium typical of modern man, the species *Homo sapiens*. The brain of these forms was about as large as that in living chimpanzees

(maximum about 600 cc), yet this organism manipulated and manufactured tools. Apparently, bipedalism freed the hands so that tool use and
construction became evolutionary possibilities in the remote ancestors of
man. Tool use by primitive near-men first appeared, as far as is known,
about 2.5 million years ago and set the stage for the origin of tool manufacture in later forms several hundred thousands years later, about 2 million years ago. Once tool manufacture was established as an attribute of
the human stock, an entirely new adaptive zone became open. Entrance
into the new zone immediately created a situation where strong natural
selection operated toward greater and greater levels of creativity in tool
manipulation and design and toward development of a larger and larger
brain to expedite control and invention of tools. Correlated with the
stimulus provided by tool manipulation and the increase in size of the
brain areas for control of these and related activities was an increase in
association centers in the brain. Very likely even in the small brains of
A. africanus the amount of brain tissue devoted to sensory and motor activity was different from that in the apes, and expansion of association
centers was probably already initiated. The gradual dominance of association centers making speech possible, increasing memory, and allowing
mental manipulation of complex and abstract symbols has given modern
man unparalleled control of his environment and himself.

As a matter of record, the principal progress in human evolution
during the preceding 500,000 years has been in the development of a
larger and more efficient brain, the expansion of tool production and design to extremely complex levels, the conscious awareness and control
of the environment through the interaction of mental and technological
activities correlated with, if not actually determined by, increasing complexity and innovation in play, ritual, language, and social organization.

The non-tool-making *Australopithecus boisei* disappeared about
1,200,000 years ago. *Australopithecus africanus* is known to have persisted
in Java and China up until 1,900,000 years ago and in South Africa until
700,000 years ago. The latter species gradually disappeared and was replaced by a more highly adapted tool-manufacturing species, *Homo erectus*,
beginning about 1,200,000 years ago. *H. erectus*, the dawn man, was originally discovered in Java by Eugene Dubois in 1894. Later finds of the
same form, often called Peking man, were made in China during the
period 1923–1936. The so-called Heidelberg man (based on a lower jaw)
found in 1907 in Germany is now also referred to this species. More recently *H. erectus* remains have been collected in Hungary, Algeria, and east
Africa. No question exists as to *H. erectus* status as a man—the species
used fire, had an advanced tool-manufacturing industry, had human morphology, and lived in a society at least as complex as that of some of the
more primitive contemporary human societies. The species differs from
modern man primarily in skull and brain development. The cranial capacity

has a maximum of 1,250 cc and the skull is primitive in several features. Early remains of this form are of an age about equal to that of the latest known *A. boisei* and overlap in time with *A. africanus.* Human types basically similar to *Homo erectus* are known to have lived up until 300,000 years ago. Apparently *A. africanus* and *H. erectus* were undergoing contemporary evolution about 2,000,000 years ago. Both forms are now extinct, and although only one species of the genus *Homo* still persists, it has become the dominant living mammal.

Members of the modern species *Homo sapiens,* to which all living men belong, first began to appear in the fossil record about 300,000 years ago. Essentially modern men in terms of cranial characters are known from 35,000 years ago. It must be noted that the course of the evolution of modern man exhibits the same pattern of divergent evolution and extinction described in previous chapters for other organisms. Although frequently cited as an example of directed evolution, the fossil record of human history demonstrates that a number of diverse lines evolved in the general direction of modern man, with a sequence of replacement of primitive forms by a more advanced type.

The success of modern man is derived from his physical and biological inheritance from his arboreal primate ancestors. Most significant to later developments were the grasping hands used for arboreal locomotion (grasping feet were lost as an adaptation to bipedal, nonarboreal locomotion) and the dependence on vision and hearing, with concurrent increased brain size. The grasping hands have made possible tool manipulation, and the large brain, originally important in sensory reception and motor response, has taken on new functions of even greater significance. Superimposed on these fundamentals was the origin of bipedal locomotion, which made subsequent evolution possible by (1) freeing the hands for tool utilization and construction and (2) uplifting the head for expansion of awareness and speech, both of which led to the rapid increase in brain size, particularly in association centers. The shift from earliest near-men to modern man in the period of 5 million years amply certifies the selective advantages gained by these latter trends.

The record of early human evolution is based, as it must be, primarily on material remains, the bones and stones of our ancestors. Nevertheless these data provide both an incomplete and, if overemphasized, an inaccurate impression of the process because they fail to provide information on the origin of those unique human features which make man an entirely new order of being, a creation of organic evolution but transcending its limits. These features involve the evolution of (1) a self-conscious human personality through increased sentience, extensive self-directed activity, lengthening of memory, and expanded awareness; (2) a human mind that transcends both empirical facts by building an ever-increasing number of concepts and symbols and the determinism of nature by providing meaning and symbolic agents of meaning that may be stored, transmitted, and translated across immense distances in time and space, and used to establish a basis for moral choice and value systems; and (3) human culture through the development of complex interpersonal and mental relationships based on ritual, language, and social organization. Man as a biological species is a unique combination of morphological, functional, ecological, and behavioral characteristics that make us distinct from any other animal. Natural man is clearly descended from earlier forms of man-like creatures with which he shares many similarities. Man stands alone in the areas of psyche, mind, and culture, each an emergent quality of human consciousness.

NATURE OF MAN

What is man?
  A multicellular animal
  A vertebrate
  A placental mammal
  A higher primate
  A biological species: *Homo sapiens*

A. *Distinctive morphological and functional features*
  Naked
  Bipedal, erect posture—with symmetrical alternating gait
  Elevated head
  Free manipulative hands, with precision grip
  Reduced masticatory apparatus (short palate, reduced lower jaw, with chin), modified dentition (reduced incisors, canines, and lower premolar teeth) and reduced chewing surfaces
  Large brain with gigantic cerebral cortex (visual and auditory centers; learning, reasoning, memory), large limbic cortex (emotions)

B. *Distinctive ecological and behavioral features*
  Dirunal, with stereoscopic color vision
  Terrestrial
  Omnivorous
  No estrus cycle
  Strong pair bonding
  Cooperative big-game hunting
  Extended dependency and care of young
  Tool using and manufacturing
  Fire using

MAN

NATURAL

EXPERIENTIAL MAN

Dreaming

Imaginative (possible, conceivable)

Symbol making and using

Talking (language)

Conceptual thinking (hierarchical, abstract, deductive, inductive, causal)

Space and form perceiving

Time binding (past, present, and future)

Social

Mystical (urge to unite with the ultimate reality both transcendent and immanent)

Ethical

Cultural

Historical—not just a product, but a molder of history

The maximum acceleration of human evolution in its cultural phases has occurred during the preceding 5000 years, that is, during the preceding 1 percent of the history of our species. Increase in mental ability, development of tool technology, and acquisition of speech have contributed to the rise of a complex sociocultural organization. Cultural evolution is nonbiological in the sense that it is independently inherited by each member of the culture regardless of his genetic origins. Each generation of mankind passes on to each succeeding generation information regarding the environment, social relations, and technology. In its earliest sense cultural evolution in human ancestors proceeded primarily by imitation. The habit of tool use was probably taught to each young *A. africanus,* and methods of tool construction were probably passed on by the same means even in the absence of speech. Acquisition of language and increased levels of mental efficiency accelerated the process of cultural inheritance; now information could be passed on verbally and even the newest methods of technology could be taught to the new generation. Advanced technological developments and extremely rapid social evolution followed on the invention of writing. Today men profit from the accumulated knowledge of thousands of generations of ancestors, with constant acceleration in improvement of techniques and ideas as the result. Cultural evolution is, of course, a reflection of our biotic inheritance; without large brains and free hands the results of our ancestors' experiences would be lost to us. However, this social inheritance of acquired knowledge proceeds at a rapid rate virtually independent of microevolutionary changes in human population.

Coupled with the rapid acceleration of cultural-intellectual evolution, man has come to move away from his rich and complicated emotional life. The heady experience of intellectual predominance and technological supremacy on the earth has led man to suppress more and more his basic human emotional and spiritual needs. Western scientific-technologic man, in particular, has developed an excessive rationality that often substitutes fantasies, ideas, and concepts generated by the intellect and mechanical tools for the reality of feelings, emotions, and association with nature and the ultimate mystery of existence. Many of modern man's greatest social and individual ills derive from his conscious denial of his basic biological and spiritual nature, which nevertheless assert themselves through unconscious motivations and lead to overt activities that distort the essence of his humanity.

The basis of human dominance of the earth, at one level of awareness, seems to be derived from our technical mastery of the environment and our cultural inheritance. From this point of view we can look forward to a period of greater social advances by the large-brained, talking, tool-making, symbol maker we call man. According to this view, in the not too distant future man may well have complete control of his earthly environment as the result of interaction between his biotic and cultural evolution, unless of course evolution in creative ideas and feelings lag too far behind our technological progress. In this framework we stand on the verge of a magnificent era in human history; our principal problem for the future is learning to live not with our environment, but with ourselves.

But from another vista, man's future role is different. He remains the unfinished ever-unfolding animal, who both feels and thinks, is both non-rational and rational, is both emotional and intellectual, is both determined and limited and indetermined and free. He is the latest creation of 15 billion years of cosmic evolution that led slowly from the level of strictly determined physical and chemical systems and reactions to the origin of life and succeedingly higher and higher levels of responsive consciousness. In this view man stands as the focal point of consciousness between his inner world and the universe. The history of life seems centered on an evolutionary pathway toward making conscious more and more of the information stored or received by an organism. If this trend continues, as it seems it must, evolution will go on until all that lies outside of awareness, that which is unconscious or unknown but present in man, becomes conscious. Man's triumphs of the last several centuries, especially under the aegis of scientific imagination, has taken the mind to the ultracosm of atomic structure and the macrocosm of outer space to create an unparalleled new conception of external reality. The future requires man to plumb his own depths (the microcosm) where distances far exceed those from earth to the outer limits of the universe and where the information content of the cosmos may be stored, to the same degree that he has explored external reality. For it is in the core of man, that unexpected constantly creating universe somehow touched by all existences and all times, that man's uncommitted creativity resides and the mystery of our existence and its ultimate relationship to the cosmos will be found. It is at this level of expanded consciousness that the next great evolutionary leap will come and man will have fulfilled his cosmic destiny.

FURTHER READING

Arieti, S., *The Intrapsychic Self*. New York: Basic Books, 1967.

Campbell, B., *Human Evolution*, 2d ed. Chicago: Aldine, 1974.

Campbell, J., *The Flight of the Wild Gander*. New York: Viking, 1969.

Eiseley, L., *The Firmament of Time*. New York: Atheneum, 1962.

Howells, W., *Evolution of the Genus Homo*. Reading, Mass.: Addison-Wesley, 1973.

Leakey, R. E. "Hominids in Africa," *American Scientist*, vol. 64 (1976), pp. 174–178.

Mumford, L., *The Myth of the Machine, Technics and Human Development*. New York: Harcourt Brace Jovanovich, 1966.

Rose, S., *The Conscious Brain*. New York: Knopf, 1973.

Stent, G., "Limits to the scientific understanding of man," *Science*, vol. 197 (1975), pp. 1052–1057.

Swanson, C. P., *The Natural History of Man*. Englewood Cliffs, N. J.: Prentice-Hall, 1973.

Thrope, W. H., *Animal Nature and Human Nature*. Garden City, N. Y.: Anchor/Doubleday, 1974.

Wheeler, J. A., "The universe as home for man," *American Scientist*, vol. 62 (1974), pp. 683–691.

Wilson, E. O., *Sociobiology, the New Synthesis*. Cambridge: Belknap Press, 1975.

Young, J. Z., *An Introduction to the Study of Man*. Oxford: Oxford University Press, 1971.

Index

Adaptation, 9, 96, 136–137; ecologic, 111

Adaptive experimentation, 144, 145; radiation, 99, 128, 136–137, 139, 147; value (W), 82

Adaptive zones, 133, 136, 137, 139, 140

Adenine, 11, 12

Aerobic respiration, 161, 162

Algae, 18

Allele, 41

Alleles, multiple, 63

Alligator mississippiensis, 79

Allopatric populations, 106

Amino acids, 10

Anaerobic respiration, 160

Anapsid, 139

Anthropoidea, 166

Apes, 170

Apomixis in speciation, 126

Archosauria, 140

Atoms, 6, 7

Australopithecus, 168, 171, 172, 174

Autocatalysis, 9, 158

Autosynthesis, 9, 158

Autotrophs, 160

Aves, 142

Aye-ayes, 166

Bacteria, 18

Binomial expansion, 45

Biological systems, new, 144

Biston betularia, 84

Blood types, human, 64

Brain, evolution and size, 169–170

Bufo, 114

Bullfrog, 114

Callithricidae, 166

Carbohydrates, 6

Carbon atoms, molecules containing, 6

Cebidae, 166

Cell, definition, 6, 7; division, 34–35; human, 33

Centromere, 35

Cercopithecidae, 166

Channelization, 148

Character cline, 112

Chiasmata, 37, 67, 68

Chimpanzee, 166

Chlorophyll, 160

Chromosome, 32, 33; duplication, 34; mutation, 60, 65

Ciliates, 18

Cline, character, 112; formation, 119

Clupea pallasi, 112

Coacervates, 157

Colloidal mixture, 7

Community, 6, 7

Competition, 128

Convergence, in special adaptation, 142

Cotylosauria, 139

Creationists, 151

Crocodiles, 142, 143

Crossing over, 37, 67–69

Crotalus viridis, 107, 108, 122

Cryptomonads, 18

Cultural evolution, 5

Cytosine, 11, 12

Darwin, C. R., 26, 97

Darwinian fitness, 82; selection, 78

Darwinism, 26

Daubentoniidae, 166

Deletion, 65

Deme, 104–105, 111, 112, 117, 118; formation, 117; in a species, 106

Demic fragmentation, 117; spatial isolation, 118

Demic variation, pattern of, 111

Dentition, 170, 171

Deoxyribonucleic acid (DNA), 157–158; mutation, 69; replication, 33, 39, 40, 48; structure, 11; virus, 7; in zygote, 40

Descent with modification, 76

Desmognathus, 136

Diapsid, 140

Diatoms, 18

Dinoflagellates, 18

Dinosaurs, 142

Diploid ($2N$) cell, 33

Directed evolution, 175

Divergence, 116, 117, 142; and breakthrough, 145; molded by natural selection, 147